T0314395

Novel bioremediation processes for treatment of seleniferous soils and sediment

Shrutika Laxmikant Wadgaonkar

Joint PhD degree in Environmental Technology

Docteur de l'Université Paris-Est

Spécialité : Science et Technique de l'Environnement

Dottore di Ricerca in Tecnologie Ambientali

Degree of Doctor in Environmental Technology

Tesi di Dottorato – Thèse – PhD thesis

Shrutika Laxmikant WADGAONKAR

Novel bioremediation processes for treatment of seleniferous soils and sediment

To be defended on December 18, 2017

In front of the PhD committee

Prof. Dr. Geoff Gadd	Reviewer
Prof. Dr. Paul Mason	Reviewer
Prof. Dr. Ir. Piet N.L. Lens	Promotor
Dr. Hab. Giovanni Esposito	Co-Promotor
Dr. Hab. Eric D. van Hullebusch	Co-Promotor
Prof. Eddy Moors	Examiner

Erasmus Joint doctoral programme in Environmental Technology for Contaminated Solids, Soils and Sediments

(ETeCoS³)

Thesis committee:

Thesis Promotor

Prof. Dr. Ir. Piet N.L. Lens
Professor of Environmental Biotechnology
UNESCO-IHE, Institute for Water Education
Delft, The Netherlands

Thesis Co-Promotors

Dr. Hab. Giovanni Esposito
Assistant Professor of Sanitary and Environmental Engineering
University of Cassino and Southern Lazio
Cassino, Italy

Dr. Hab. Eric D. van Hullebusch
Hab. Associate Professor in Biogeochemistry
University of Paris-Est
Marne-la-Vallée, France

Other Members

Prof. Dr. Geoff Gadd
College of Life Sciences,
University of Dundee,
Dundee, Scotland, UK

Prof. Dr. Paul Mason
Department of Geosciences
Utrecht University
Utrecht, The Netherlands

This research was conducted under the auspices of the Erasmus Mundus Joint Doctorate Environmental Technologies for Contaminated Solids, Soils, and Sediments (ETeCoS3) and the Graduate School for Socio-Economic and Natural Sciences of the Environment (SENSE)

Published by:
CRC Press/Balkema
Schipholweg 107C, 2316 XC, Leiden, the Netherlands
Pub.NL@taylorandfrancis.com
www.crcpress.com – www.taylorandfrancis.com
ISBN: 978-1-138-38480-4

Table of Contents

Acknowledgement

First, I would like to thank Erasmus Mundus ETeCoS3 program (FPA no. 2010-0009) for providing the financial support to carry out this thesis. I would also like to thank UNESCO - IHE, University of Paris-Est, University of Cassino and Southern Lazio, Helmholtz-Centre for Environmental Research – UfZ and University of Naples Federico II for hosting me.

I hereby express my sincere gratitude to Prof. Piet Lens (UNESCO-IHE, The Netherlands), promotor of the thesis, for his encouragement, support and constant supervision of the progress of my research. His scientific suggestions and constructive comments on the research was critical to achieve success. I would also like to thank Dr. Giovanni Esposito (University of Cassino and the Southern Lazio, Italy) co-promotor of the thesis for his support, encouragement and insightful comments in improving the quality of the thesis along with constant support on the administrative affairs during this PhD. I also would like to thank Dr. Eric D. van Hullebusch (University of Paris-Est, France) for his insightful comments and suggestions during the scientific meetings and summer schools along with the administrative support during this PhD. I am extremely grateful to Dr. Y. V. Nancharaiah (Venkata), Bhabha Atomic Research Centre, Kalpakkam, India. His extensive knowledge in the fields of bioremediation, bioreactors and microbiology indeed helped me shape this thesis. His constant support during experiments and proficient comments on the manuscripts were indispensable for this research. and many thanks to Dr. Eldon to always be there in the time of need and for your constant support.

I am grateful to Prof. Karaj S. Dhillon (Punjab Agricultural University, Ludhiana, India) for providing me with the seleniferous soil, without which this research was not possible. I would also like to thank Dr. Carsten Vogt, Dr. Ivonne Nijenhuis and Dr. Niculina Musat for welcoming me to Helmholtz-Centre for Environmental Research – UfZ and for all their guidance and support during the mobility. I would like to acknowledge the support from the COST action ES1302 (Ref. No. COST-STSM-ES1302-33921) at Universität des Saarlandes (Germany). I would also like to thank Dr. Jacob Claus, Jawad, Ammar, Yaman and Reem for welcoming me to University of Saarland for their guidance and support during the Short-term scientific mission. I thank Prof. Patrick Chaimbault of the Université de Lorraine (France) for for his technical assistance and expertise. I am immensely thankful to Prof. Massimiliano Fabbricino, Dr. Alberto Ferraro, Dr. Marco Race and Dr. Ludovico Pontoni for welcoming me to University of Napoli, Federico II along with their constant guidance and support in experiments as well as manuscript preparation during my mobility to Italy.

A very special thanks to Prof. Geoff Gadd (University of Dundee), and Dr. Paul Mason (Utrecht University), jury members, for their critical comments and suggestions which led me thinking more deeply about my research and had provided me with a new perspective of my thesis. Moreover, it has also given me thoughts on how to improve this work further. I would like to express my gratitude towards Prof. Eddy Moors for chairing my PhD defence session.

My gratitude also goes to IHE-lab staff, Fred, Frank, Ferdi, Peter, Berend and Lyzette for your support and maintenance of the laboratory instruments in perfect conditions. I realised how important you guys are during my mobility to other laboratories. Thank you, Frank, for your constant support with the GF-AAS and GC. Thank you, Peter, for all your help with the microbiology and thank you Lyzette for helping me work with the IC. Thank you Ferdi and Berend for teaching me how to use the electron microscope. I thank you all for the training and support in the lab.

This section would be incomplete without acknowledging my colleagues and now friends Lea, Tejaswini, Samayita, Joy, Chiara, Susma, Carlos, Bijit, Nirakar, Anna, Feishu, Alessandra, Chris, Iosif, Mirko, Viviana, Angelo, Gabriele, Bikash, Jairo and Sofia. It has been a pleasure knowing you and working with you all.

Finally, I would like to express my deepest gratitude to my family for their blessings and love. Thanks to my parents for supporting me to chase my dreams despite the distance and time. Without your constant support, this would not have been possible. This section will not be complete without thanking Neeraj for his encouragement and understanding. You have always been supportive and to someone that I can look upto and hope to keep knowing you for the rest of my life.

Summary

The aim of this Ph.D. was to develop a technology for the remediation of seleniferous soils/sediments and to explore microbial reduction of selenium oxyanions under different respiration conditions and bioreactor configurations.

Seleniferous soil collected from the wheat-grown agricultural land in Punjab (India) was characterized and its soil washing was optimized by varying parameters such as reaction time, temperature, pH and liquid to solid ratio. In order to maximize selenium removal and recovery from this soil, effect of competing ions and oxidizing agents as chemical extractants for soil washing were also studied. Although oxidizing agents showed a maximum selenium removal efficiency (39%), the presence of oxidizing agents in the leachate and the agricultural soil may increase the cost of their post-treatment. Aquatic plants, *Lemna minor* and *Egeria densa* were used to study phytoremediation of the soil leachate containing oxidizing agents. However, the selenium removal efficiency by aquatic weeds was significantly affected by the high concentrations of these oxidizing agents in the soil leachate.

Seleniferous soil flushing revealed the selenium migration pattern across the soil column. Migration of soluble selenium fraction from the upper to the lower layers and its subsequent reduction and accumulation in the lower layers of the soil column was observed during soil flushing. The selenium removal efficiency by the soil flushing method decreased with an increase in the column height. Furthermore, the soil leachate containing selenium oxyanions obtained from soil washing was treated in a UASB reactor by varying the organic feed. Effluent containing less than 5 μg L^{-1} selenium was achieved, which is in accordance with the USEPA guidelines for selenium wastewater discharge limit.

Moreover, *ex situ* bioremediation of selenium oxyanions was studied under variable conditions. An aerobic bacterium (*Delftia lacustris*) capable of transforming selenate and selenite to elemental selenium, but also to hitherto unknown soluble selenium ester compounds was serendipitously isolated and characterized. Alternatively, anaerobic bioreduction of selenate coupled to methane as electron donor was investigated in serum bottles and a biotrickling filter using marine sediment as inoculum. Finally, the effect of contamination of other chalcogen oxyanions in addition to selenium was studied. Simultaneous reduction of selenite and tellurite by a mixed microbial consortium along with the retention of biogenic Se and Te nanostructures in the EPS was achieved during a 120-day UASB bioreactor operation.

Abstract

Deze Ph.D. focuste zich op het ontwikkelen van een technologie voor de sanering van seleniferous bodem/sedimenten, en het bestuderen van de microbiele vermindering van selenium oxyanions onder verscheidene omstandigheden omtrent beademing en bioreactor configuraties.

Seleniferous grond, die verzameld werd uit het agrarische landschap in Punjab (India) waar voornamelijk graan wordt geteeld, werd gekarakteriseerd en de bodemspoeling werd geoptimaliseerd door verscheidene parameters, zoals reactietijd, temperatuur, pH en vloeibare-naar vaste stadium ratio. Om de verwijdering van selenium en het herstel van de grond te maximaliseren werden concurrerende ions en oxiderende middelen, zoals chemische extractiemiddelen voor bodemspoeling, bestudeerd. Alhoewel de oxiderende middelen een maximale selenium verwijderingsefficiëntie aantoonden (39%), kon de aanwezigheid van oxiderende middelen in het percolaat en de agrarische grond toch de kosten van het nabehandelingsproces verhogen. De aquatische planten, *Lemna minor* en *Egeria densa* werden gebruikt om fytoremediatie van het bodempercolaat, bestaande uit oxiderende middelen, te onderzoeken. De selenium verwijderingsefficiëntie bij aquatisch wier daarentegen, was aanzienlijk aangetast door de hoge concentraties van deze oxiderende middelen in het bodempercolaat.

Seleniferous bodemspoeling toonde het selenium migratiepatroon over de bodemkolom aan. Tevens werd er een migratie van oplosbare selenium fractie vanuit de bovenste laag naar de onderste laag, en de opeenvolgende afname en accumulatie in de lagere lagen van de bodemkolom waargenomen tijdens de bodemspoeling. De selenium verwijderingsefficiëntie door de bodemspoeling-methode nam af met een toename in de kolomhoogte. Daarnaast werd het bodempercolaat, met selenium oxyanions afkomstig van bodemspoeling, behandeld in een UASB reactor door de organische voeding te variëren. Afvalwater bestaande uit minder dan 5 μg L^{-1} selenium werd hieruit verkregen, wat overeenkomt met de USEPA richtlijnen voor het afloslimiet van selenium afvalwater.

Bovendien werd *ex situ* biosanering van selenium oxyanions bestudeerd onder variabele omstandigheden. Een aerobic bacterie (*Delftia lacustris*), die in staat is om selenaat en seleniet in elementaire selenium, en, tot dusver bekend, ook in onbekende oplosbare selenium ester verbindingen te transformeren, werd geïsoleerd en gekarakteriseerd. Anderzijds werd de anaerobic bio-afname van selenaat, gekoppeld aan methaan als elektron donor, onderzocht in serum flessen en een bio-besproeiingsfilter, waarbij zee-sediment gebruikt werd als inoculum. Tenslotte werd, naast selenium, het effect van de vervuiling van andere chalcogen oxyanions onderzocht. De gelijktijdige afname van seleniet en telluriet, die ontstond door een mix van microbieel consortium samen met het behoud van biogenische Se en Te nanostructuren in de EPS, werd bereikt tijdens een proces van de UASB bioreactor gedurende 120 dagen.

Résumé

L'objectif de cette thèse a été de développer une technologie pour l'assainissement des sols / sédiments sélénifères et d'étudier la réduction microbienne des oxy-anions de sélénium dans différentes conditions de respiration et de configurations du bioréacteur.

Le sol sélénifère prélevé, dans les terres agricoles cultivées de blé au Pendjab (Inde), a été caractérisé et son lavage a été optimisé en faisant varier les paramètres tels que le temps de réaction, la température, le pH et le rapport liquide / solide. Afin de maximiser l'élimination et la récupération du sélénium à partir de ce sol, l'effet des ions compétiteurs et les composés oxydants comme les agents d'extraction pour le lavage du sol, ont également été étudiés. Bien que les agents oxydants aient montré une efficacité maximale d'élimination du sélénium (39%), la présence d'agents oxydants dans le lixiviat et le sol agricole peut augmenter le coût de leur post-traitement. Les plantes aquatiques, *Lemma minor* et *Egeria densa* ont été utilisées pour étudier la phyto-remédiation du lixiviat du sol contenant des agents oxydants. Cependant, l'efficacité d'élimination du sélénium par les plantes aquatiques a été significativement affectée par les fortes concentrations de ces agents oxydants dans le lixiviat du sol.

Le rinçage du sol sélénifère a révélé un motif de migration du sélénium à travers la colonne du sol. La migration de la fraction de sélénium soluble de la couche supérieure vers la couche inférieure et sa réduction et son accumulation subséquentes dans les couches inférieures de la colonne de sol, ont été observées pendant le rinçage du sol. L'efficacité d'élimination du sélénium par la méthode de rinçage du sol a diminué avec une augmentation de la hauteur de la colonne. De plus, le lixiviat contenant des oxy-anions de sélénium obtenus à partir du lavage du sol, a été traité dans un réacteur UASB en faisant varier l'alimentation organique. Des effluents contenant moins de 5 μg de sélénium L-1 ont été obtenus, ce qui est conforme aux normes de l'USEPA pour la limite de rejet de sélénium dans les eaux usées.

De plus, la bio-remédiation *ex situ* des oxy-anions de sélénium a été étudiée dans des conditions variables. Une bactérie aérobie (*Delftia lacustris*) capable de transformer le sélénate et le sélénite en sélénium élémentaire, mais aussi en composés d'ester de sélénium solubles jusque-là inconnus, a été isolée et caractérisée de manière fortuite. Alternativement, la bio-réduction anaérobie du sélénate couplé au méthane en tant que donneur d'électrons, a été étudiée dans des bouteilles de sérum et un filtre percolateur en utilisant des sédiments marins comme inoculum. Enfin, l'effet de la contamination d'autres oxy-anions chalcogènes, en plus du sélénium, a été étudié. La réduction simultanée de la sélénite et de la tellurite par un consortium microbien mixte ainsi que la rétention des nanostructures de Se et de Te biogènes dans l'EPS, ont été réalisées durant une opération de 120 jours dans un bioréacteur UASB.

Sommario

Nel presente dottorato di ricerca, l'obiettivo principale è stato incentrato sullo sviluppo di un processo atto alla bonifica di suoli/sedimenti seleniferi oltre che sullo studio del fenomeno di riduzione, per via biologica, degli ossoanioni del selenio, investigato in differenti condizioni di respirazione microbica e configurazioni reattoristiche.

I campioni di suolo selenifero, raccolti da un'area agricola finalizzata alla coltivazione del grano nello stato di Punjab (India), sono stati analizzati per determinarne le caratteristiche fisico-chimiche. Inoltre, esperimenti volti all'ottimizzazione di una procedura di 'soil washing', impiegata come tecnica di bonifica dei campioni raccolti, sono stati condotti al variare di differenti parametri di processo quali: tempo di trattamento, temperatura di processo, pH e rapporto liquido-solido. Al fine di massimizzare le efficienze di rimozione e recupero del selenio dal suolo oggetto di studio, sono state condotte ulteriori prove sperimentali di 'soil washing' impiegando ioni competitivi e agenti ossidanti nelle soluzioni chimiche di lavaggio. I risultati ottenuti hanno mostrato massime efficienze di rimozione del selenio, pari al 39%, conseguibili mediante l'utilizzo di agenti ossidanti; nonostante ciò, l'impiego dei suddetti agenti chimici, rappresenterebbe un possibile fattore d'incremento dei costi legati al post-trattamento delle soluzioni di lavaggio, contenenti gli agenti ossidanti, e del suolo agricolo soggetti al processo di 'soil washing'. All'uopo, prove di fitodepurazione, per il trattamento della soluzione di lavaggio contenente agenti ossidanti, sono state condotte mediante l'impiego di piante acquatiche quali la *Lemna minor* e *Egeria densa*. Ad ogni modo, tali prove, hanno evidenziato che l'efficienza di rimozione del selenio dalla soluzione di lavaggio, ad opera delle suddette piante acquatiche, risulta fortemente influenzata dalla presenza degli agenti ossidanti impiegati.

Ulteriori prove sperimentali sono state condotte simulando un processo di 'soil flushing', sui campioni di suolo selenifero, al fine di determinare i profili di lisciviazione del selenio all'interno della colonna di suolo. Dai risultati, è stato possibile osservare la migrazione della frazione solubile del selenio dagli strati superiori della colonna a quelli inferiori, in cui si sono verificati conseguenti fenomeni di riduzione e accumulo di selenio durante il decorrere della prova. In questo caso, l'efficienza di rimozione del selenio è risultata correlata all'altezza della colonna di suolo impiegata, ottenendo rendimenti di rimozione decrescenti al crescere dell'altezza della colonna.

Un successivo step della presente ricerca è stato focalizzato sul trattamento della soluzione di lavaggio contenente ossoanioni di selenio, ottenuta a valle del processo di 'soil washing', all'interno di un reattore UASB esercito facendo variare il carico organico in ingresso. Le analisi sull'effluente, ottenuto a valle del trattamento mediante reattore UASB, hanno mostrato concentrazioni di selenio inferiori ai 5 μg L^{-1}; tali valori rientrano nei limiti ambientali previsti per lo scarico delle acque reflue contenenti selenio suggeriti dalle linee guida dell'agenzia USEPA.

Prove volte a simulare processi *ex situ* di biorisanamento di ossoanioni di selenio sono state condotte al variare di differenti condizioni. In tali prove, è stato isolato e caratterizzato un batterio aerobico, il *Delftia lacustris*, in grado di convertire selenati e seleniti in selenio elementare e, inoltre, in esteri di selenio solubile ancora non identificati. Altre prove, invece, volte allo studio della riduzione biologica, in condizioni anaerobiche, del selenato accoppiato al metano usato come elettron-donatore, sono state condotte in bottiglie per colture anaerobiche e tramite letto percolatore impiegando sedimenti marini in qualità di inoculo. In conclusione, oltre al selenio, è stato investigato l'effetto della contaminazione dovuta ad ulteriori ossoanioni calcogeni. I risultati hanno mostrato il verificarsi di un simultaneo fenomeno di riduzione del selenito e tellurito, dovuto all'azione di un consorzio microbico misto, e dell'accumulo di nanostrutture di selenio e tellurio biogenico all'interno dell'EPS durante l'esercizio di 120 giorni del reattore UASB.

CHAPTER 1

General Introduction

1.1. Background

Selenium (Se) is a redox-sensitive trace element that belongs to the chalcogen family (Group 16) of the periodic table and exhibits properties similar to that of sulfur (Keskinen et al., 2009). It is an essential element for animals and humans and plays an important role in redox regulation of intracellular signalling, redox homeostasis and thyroid hormone metabolism (El-Ramady et al., 2015). In humans and animals, dietary uptake of selenium lower than 40 μg day^{-1} causes selenium deficiency, which leads to multiple sclerosis, muscular dystrophy, heart disease, immune system disorders cancer and reproductive disorder (El-Ramady et al., 2015). The World Health Organization (WHO) recommended the dietary allowance of 70 - 350 μg day^{-1} selenium for humans. Dietary uptake higher than 400 μg day^{-1} leads to selenium toxicity which results in hair and nail loss and disruption of the nervous and digestive systems in humans and animals (Lenz et al., 2009). It is this narrow threshold of the recommended dietary allowance that makes this study of selenium in the environment essential.

The amount of selenium in the food chain and thus in human body depends on the selenium soil content, which varies greatly throughout the world. Therefore, soil is the most important part of the environment where determination of selenium is necessary (Hagarova et al., 2003). The common range of Se in soils is 0.01 - 2 mg kg^{-1}, but the distribution varies from almost zero to 1250 mg kg^{-1} (Oldfield, 2002). The occurrence of selenium in soil has different sources such as lithogenic, pedogenic, atmospheric, phytogenic, and anthropogenic (Kabata-pendias et al., 2001). About 37 - 40% of the total selenium emissions to the atmosphere are due to the anthropogenic activities such as Se applied to soils, foliar sprays, seed treatments, and phosphatic fertilizers for agriculture and fly-ash, smelting or washing of Se-rich ores and some sewage sludges from industries (El-Ramady et al., 2015).

1.2. Problem Statement

Selenium content in soil varies greatly throughout the world. Higher amounts of bioavailable forms of selenium in soils greatly influence the amount of selenium in the food chain (**Figure 1.1**). Elevated selenium content in soil can lead to contamination of water bodies and ground waters due to the leaching caused by rainfall and irrigation (Wu, 2004). Quantification of total selenium does not actually give information about the chemical species, thus giving no information about its bioavailability to plants. Various forms of selenium are soluble, exchangeable, bound to organic matter, sulfides, carbonates and oxides. Selenium occurs mainly in four oxidation states in soil, viz. selenate(VI), selenite(IV), elemental selenium(0) and selenide(-II). The chemical forms of selenium and their solubility depends on the redox potential and pH of the soil. Other factors that contribute to selenium speciation are organic matter content, iron oxide levels, clay type and content. The selenium available to the plants is influenced by several soil factors such as pH, salinity and calcium carbonate content (Hagarova et al., 2003).

Figure 1.1. Wheat grown agricultural soil with high selenium content from northwest India adversely affects the productivity of the crops and leads to bioaccumulation and biotransfer of selenium to higher trophic levels

In situ bioremediation approaches like phytoremediation using selenium hyper-accumulating plants (Bañuelos et al., 2015) have proved promising for removing selenium from soil using plants. The application of these selenium rich plants as selenium supplements in selenium deficient populations or fertilizer for selenium deficient agricultural soils is being studied (Yasin et al., 2014). *In situ* selenium volatilization (Dhillon et al., 2010; Flury et al., 1997) and dissimilatory selenate reduction (Oremland et al., 1991; Fellowes et al., 2013) using soil amendment with organic carbon and/or protein have been studied for biotreatment of seleniferous soils (Calderone et al., 1990; Frankenberger and Arshad, 2001). Since soil amendment approaches show no statistically significant difference in the selenium volatilization or dissimilatory reduction from those without soil amendment (Flury et al., 1997), additional studies are warranted to apply *in situ* selenium bioremediation strategies.

Although *in situ* bioremediation has a potential to provide a low cost and efficient technology for remediation of seleniferous soils, the process is relatively slow, thus increasing the risk of groundwater contamination due to leaching of selenium oxyanions and re-oxidation of the reduced selenium compounds. Therefore, it is necessary to explore other technologies for remediation of seleniferous soils.

Recent studies showed that selenium contamination of marine environments due to effluent discharges from mining industries or coal fired power plants is a major environmental concern (Ellwood et al., 2016; Laird et al., 2014). Adverse effects on marine flora and fauna due to bioaccumulation and biotransfer of selenium to higher trophic levels was observed (Turner, 2013). A significant flux of selenium from the sediment to marine water was also recorded (Meseck and Cutter, 2012). In addition, selenium anthropogenic activities such as mining and refinery industries can contaminate soil-water environments with Se and Te oxyanions (Jorgenson, 2002; Perkins, 2011). Therefore, it is essential to explore novel bioremediation processes to consider these environmental concerns.

Selenate and selenite reduction under anaerobic conditions has been studied for industrial wastewaters (Lenz et al., 2008; Tan et al., 2016) and found to be a promising approach for treatment of selenium bearing wastewaters. *Ex situ* remediation technologies have been applied to the remove heavy metals from polluted soils. Similar strategies involving leaching of Se from soil and treatment of leachate in bioreactors may allow remediation of Se-contaminated or seleniferous soils. The *ex situ* remediation of seleniferous soils ensures prevention of contamination to other natural resources as it completely removes Se from contaminated soils. The *ex situ* approach not only removes selenium from the soil, thus cleaning it, but also assists in recovery and recycling this critical and scarce element to meet its increasing industrial demand (Nancharaiah et al., 2016).

1.3. Research objectives

The aim of this research is to develop technologies for *ex situ* remediation of seleniferous soils by combining soil washing and biotreatment of the soil leachate under varying conditions.

The specific objectives are:

1. Optimisation of soil washing procedure
   a. To investigate physical and chemical characteristics of seleniferous soils.
   b. To optimize removal of selenium from the soil by varying treatment parameters.
   c. To study the effect of competing ions and oxidising agents on selenium removal from soil.
2. Biological treatment of seleniferous soil and soil leachate

a. To investigate the effect of soil amendments with different electron donors and bioaugmentation on selenate reduction in soil.

b. Bioremediation of soil leachate using anaerobic granular sludge in a UASB bioreactor.

c. Phytoremediation of soil leachate using aquatic plants, *Lemma minor* and *Egeria densa*.

3. Biological treatment of artificially simulated seleniferous soil leachate under varying environmental conditions

   1.1. To investigate aerobic reduction of selenium oxyanions.

   1.2. To investigate anaerobic selenate reduction coupled to methane oxidation by marine lake sediment.

   1.3. To study simultaneous bioreduction of selenite and tellurite from artificially simulated seleniferous soil leachate in a UASB bioreactor.

1.4. Structure of thesis

This dissertation comprises of nine chapters. An overview of the structure of the dissertation is given in **Figure 1.2**. A brief outline of each chapter is as follows:

Chapter 1 presents a general overview of the thesis, including background, problem description, research objectives and thesis structure. Chapter 2 provides the literature review about the impact of seleniferous soil on the environment and an overview of the current technologies undertaken for bioremediation of seleniferous soils. The *ex situ* technique for soil bioremediation has been suggested and research gaps in the field have been highlighted.

Chapter 3 explores the possibility of *ex situ* remediation of a seleniferous soil sampled from Punjab (India) using soil washing techniques. Chapter 4 combines *in situ* and *ex situ* methods for soil remediation, where the seleniferous soil was treated in an *in situ* microcosm experiment, while the soil leachate optimally produced from experiments in chapter 3 was treated *ex situ* in a UASB bioreactor. Chapter 5 evaluated the potential of phytoremediation of seleniferous soil leachate containing oxidising agents and competing ions using the aquatic plants *Lemna minor* and *Egeria densa*.

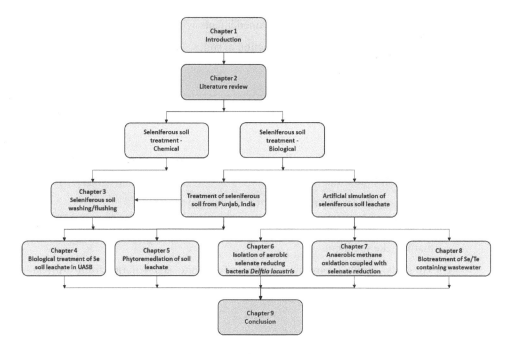

Figure 1.2. Overview of the chapters in this PhD thesis

Chapter 6 describes isolation and characterization of a novel aerobic bacterium, *Delftia lacustris*, capable of selenate and selenite reduction to elemental selenium and soluble selenium ester compounds. Chapter 7 explores a novel mechanism for selenate reduction coupled to methane oxidation by anaerobic marine sediment in batch enrichments and in a biotrickling filter. Chapter 8 shows the possibility of treating artificial soil leachate containing Se and Te oxyanions in a UASB bioreactor. Simultaneous reduction of selenite and tellurite by anaerobic granular sludge was associated with the formation of biogenic Se(0), Te(0) and Se-Te nanostructures.

Chapter 9 summarizes and draws conclusions on the knowledge gained from the above studies. It also gives recommendations and prospects for future research.

References

Bañuelos, G.S., Arroyo, I., Pickering, I.J., Yang, S.I., Freeman, J.L., 2015. Selenium biofortification of broccoli and carrots grown in soil amended with Se-enriched hyperaccumulator *Stanleya pinnata*. Food Chem. 166, 603–8.

Calderone, S.J., Frankenberger, W.T., Parker, D.R., Karlson, U., 1990. Influence of temperature and organic amendments on the mobilization of selenium in sediments. Soil

Biol. Biochem. 22, 615–620.

Dhillon, K.S., Dhillon, S.K., Dogra, R., 2010. Selenium accumulation by forage and grain crops and volatilization from seleniferous soils amended with different organic materials. Chemosphere 78, 548–56.

El-Ramady, H., Abdalla, N., Alshaal, T., Domokos-Szabolcsy, É., Elhawat, N., Prokisch, J., Sztrik, A., Fári, M., El-Marsafawy, S., Shams, M.S., 2015. Selenium in soils under climate change, implication for human health. Environ. Chem. Lett. 13, 1.

Ellwood, M.J., Schneider, L., Potts, J., Batley, G.E., Floyd, J., Maher, W.A., 2016. Volatile selenium fluxes from selenium-contaminated sediments in an Australian coastal lake. Environ. Chem. 13, 68–75.

Fellowes, J.W., Pattrick, R.A.D., Boothman, C., Al Lawati, W.M.M., van Dongen, B.E., Charnock, J.M., Lloyd, J.R., Pearce, C.I., 2013. Microbial selenium transformations in seleniferous soils. Eur. J. Soil Sci. 64, 629–638.

Flury, M., Jr, W.T.F., Jury, W.A., 1997. Long-term depletion of selenium from Kesterson dewatered sediments 198, 259–270.

Frankenberger, W.T., Arshad, M., 2001. Bioremediation of selenium-contaminated sediments and water. BioFactors 14, 241–254.

Hagarova, I., Zemberyova, M., Bajcan, D., 2003. Sequential and single step extraction procedures for fractionation of selenium in soil samples. Chem. Pap. 59, 93–98.

Jorgenson, J.D., 2002. Selenium and tellurium, in: U.S. Geological Suvery Minerals Yearbook. pp. 1–7.

Kabata-pendias, A., Pendias, H., 2001. Trace elements in soils and plants trace elements in soils and plants, 3rd ed. CRC press LLC, Boca Raton, Florida.

Keskinen, R., Ekholm, P., Yli-Halla, M., Hartikainen, H., 2009. Efficiency of different methods in extracting selenium from agricultural soils of Finland. Geoderma 153, 87–93.

Laird, K.R., Das, B., Cumming, B.F., 2014. Enrichment of uranium, arsenic, molybdenum, and selenium in sediment cores from boreal lakes adjacent to northern Saskatchewan uranium mines. Lake Reserv. Manag. 30, 344–357.

Lenz, M., Hullebusch, E.D. Van, Hommes, G., Corvini, P.F.X., Lens, P.N.L., 2008. Selenate removal in methanogenic and sulfate-reducing upflow anaerobic sludge bed reactors. Water Res. 42, 2184–94.

Lenz, M., Lens, P.N.L., 2009. The essential toxin: the changing perception of selenium in environmental sciences. Sci. Total Environ. 407, 3620–33.

Meseck, S., Cutter, G., 2012. Selenium behavior in San Francisco bay sediments. Estuaries and

Coasts 35, 646–657.

Nancharaiah, Y. V, Mohan, S.V., Lens, P.N.L., 2016. Biological and bioelectrochemical recovery of critical and scarce metals. Trends Biotechnol. 34, 137–155.

Oldfield, J.E., 2002. Se World Atlas. Selenium-tellurium development association, Grimbergen, Belgium.

Perkins, W.T., 2011. Extreme selenium and tellurium contamination in soils - An eighty year-old industrial legacy surrounding a Ni refinery in the Swansea Valley. Sci. Total Environ. 412–413, 162–169.

Roland S. Oremland, Nisan A. Steinberg, Theresa S. Presser, L.G.M., 1991. *In situ* bacterial selenate reduction in agricultural drainage systems in western Nevada. Appl. Environ. Microbiol. 57, 615–617.

Tan, L.C., Nancharaiah, Y. V, van Hullebusch, E.D., Lens, P.N.L., 2016. Selenium: Environmental significance, pollution, and biological treatment technologies. Biotechnol. Adv. 34, 886–907.

Turner, A., 2013. Selenium in sediments and biota from estuaries of southwest England. Mar. Pollut. Bull. 73, 192–198.

Wu, L., 2004. Review of 15 years of research on ecotoxicology and remediation of land contaminated by agricultural drainage sediment rich in selenium. Ecotoxicol. Environ. Saf. 57, 257–69.

Yasin, M., El Mehdawi, A.F., Jahn, C.E., Anwar, A., Turner, M.F.S., Faisal, M., Pilon-Smits, E. A. H., 2014. Seleniferous soils as a source for production of selenium-enriched foods and potential of bacteria to enhance plant selenium uptake. Plant Soil 386, 385–3

CHAPTER 2

Literature review - Environmental impact and bioremediation of seleniferous soils and sediments

This chapter has been modified and published as:

Wadgaonkar SL, Nancharaiah YV, Esposito G, Lens PNL (2018) Environmental impact and bioremediation of seleniferous soils and sediments. Crit. Rev. Biotechnol. 38(6): 941-956. DOI: 10.1080/07388551.2017.1420623

Abstract

Selenium concentrations in the soil environment are directly linked to its transfer in the food chain, eventually causing either deficiency or toxicity associated with several physiological dysfunctions in animals and humans. Selenium bioavailability depends on its speciation in the soil environment, which is mainly influenced by prevailing pH, redox potential, and organic matter content of the soil. The selenium cycle in the environment is primarily mediated through chemical and biological selenium transformations. Interactions of selenium with microorganisms and plants in the soil environment have been studied in order to understand the underlying interplay of selenium conversions and to develop environmental technologies for efficient bioremediation of seleniferous soils. *In situ* approaches such as phytoremediation, soil amendment with organic matter and biovolatilization are promising for remediation of seleniferous soils. *Ex situ* remediation of contaminated soils by soil washing with benign leaching agents is widely considered for removing heavy metal pollutants. However, it has not been applied until now for remediation of seleniferous soils. Washing of seleniferous soils with benign leaching agents and further treatment of Se-bearing leachates in bioreactors through microbial reduction will be advantageous as it is aimed at removal as well as recovery of selenium for potential re-use for agricultural and industrial applications. This review summarizes the impact of selenium deficiency and toxicity on ecosystem in, respectively, selenium deficient and seleniferous regions across the globe and recent research in the field of bioremediation of seleniferous soils.

Keywords: ex situ bioremediation, in situ bioremediation, phytoremediation, selenium bioreduction, selenium bioremediation, selenium fortification, seleniferous soil, Se toxicity, soil washing.

2.1. Introduction

Selenium (Se $_{79}^{34}$), named after the Greek goddess of the moon "Selene", was discovered in 1817 by the Swedish chemist Jöns Jacob Berzelius (Lenz et al., 2009). This redox-active element is a metalloid positioned between sulfur and tellurium in the group 16 (chalcogen family) and between arsenic and bromine in period 4 of the periodic table (Keskinen et al., 2009). Relatively low in abundance, selenium is ranked 69th in order of abundance in earth's crust (Dhillon and Dhillon, 2003).

On a commercial scale, selenium has a wide range of applications in electronics, solar cells, glass industry, photocopying, cosmetic industries (Naumov 2010) and medicine as dietary supplement (Bodnar et al., 2016). Therefore, it is considered as a high commercial value element. In agriculture, selenium has been applied as fertiliser in selenium deficient regions (Lavu et al., 2012) and has also been used to inhibit the formation of sulfate-mediated methylmercury in paddy fields, wherein amendment of selenium in soil leads to the formation of stable and non-toxic Hg-Se complex(es) (Wang et al., 2016). Being an essential element, selenium intake is recommended via food supplements in order to avoid deficiency in humans and animals (Rautiainen et al., 2016).

While recovery of selenium from anode slime formed as a by-product of copper refining (Nancharaiah et al., 2016) may prove insufficient to meet the industrial demand in the future, there is a need to reuse and recover selenium from the soil-water environment using novel means to replenish the depleting selenium resources. Several soil-water environmental settings across the globe contain excess of selenium (Oldfield, 2002) that may be detrimental to the local flora and fauna. Recovery of selenium from these Se-rich environments would nevertheless prove to be beneficial, both ecologically and economically.

Research on selenium cycling across the soil-plant-atmosphere (Winkel et al., 2015), role of biofortification on improving nutritional quality of crops (Malagoli et al., 2015) and *Brassicaceous* vegetables (Wiesner-Reinhold et al., 2017), biochemistry of selenium toxicity and accumulation in plants (Gupta and Gupta, 2017; Schiavon and Pilon-Smits, 2017), selenium enrichment in horticultural crops (Puccinelli et al., 2017), microbiology of selenium respiring bacteria (Nancharaiah and Lens, 2015a), biotechnological applications of selenium biomineralization (Nancharaiah and Lens, 2015b) and biological treatment of selenium wastewaters (Tan et al., 2016) have been recently reviewed. In this review, we aim to provide

11

an up-to-date overview of research on the selenium cycle in the context of bioremediation of seleniferous soils and sediments. **Figure 2.1** provides a general overview of seleniferous soil contamination and *in situ* bioremediation techniques applied. The research performed on the natural and anthropogenic sources of selenium in soils and the adverse effects of elevated soil-Se content on the environment is discussed. Particular attention is paid to discuss the cause and effect of selenium toxicity on flora and fauna associated with seleniferous soils and sediments as well as technologies for bioremediation of seleniferous soils coupled to selenium recovery.

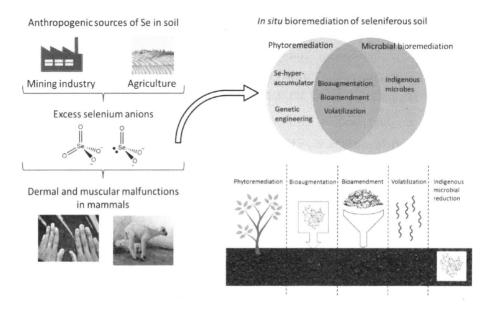

Figure 2.1. General schematic of seleniferous soil causes, effect and bioremediation

2.2. Selenium in the soil environment

2.2.1. Selenium content and species present in soils and sediments

Selenium concentrations in soil vary widely throughout the world. Although the normal range of selenium in soil ranges between 0.1 to 2 μg Se g^{-1}, the selenium content in soil varies from 0 to 100 μg Se g^{-1} throughout the world (Oldfield, 2002) and can reach up to 1200 μg Se g^{-1} in selenium rich soils at a site near County Meath, Ireland (Winkel et al., 2012). Soils containing less than 0.05 μg Se g^{-1} are considered selenium-deficient, whereas those containing more than 5 μg Se g^{-1} soil are termed seleniferous soils (Oldfield, 2002). Distribution of selenium deficient and seleniferous regions across the world is shown in **Figure 1**. Non uniform distribution of selenium in the Earth's crust creates both selenium deficient and seleniferous soils adjacent to each other, sometimes separated by a distance of only 20 km (Fordyce, 2007) (**Figure 2.2**).

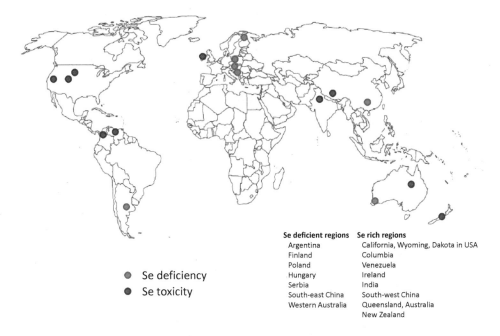

Se deficient regions   Se rich regions
Argentina   California, Wyoming, Dakota in USA
Finland   Columbia
Poland   Venezuela
● Se deficiency   Hungary   Ireland
                        Serbia   India
● Se toxicity   South-east China   South-west China
                        Western Australia   Queensland, Australia
                                                New Zealand

Figure 2.2. Map showing the prominent regions having selenium in the world. Selenium rich areas are marked red; selenium deficient areas are marked blue (Fordyce, 2013; Kaur et al., 2014; Oldfield, 2002)

In soil, selenium is present in both inorganic and organic forms (Goh and Lim, 2004). About 50 minerals are formed by the combination of selenium with metals in soil (Kabata-pendias et al., 2001). Commonly found inorganic forms of selenium in soil are klockmannite (CuSe), ferroselite ($FeSe_2$), clausthalite (PbSe), naumannite (Ag_2Se) and tiemannite (HgSe) (**Figure 2.3**). The majority of organic forms of selenium present in soil are analogues of sulfur compounds (Kabata-pendias et al., 2001). The major geochemical characteristic exhibited by selenium is its chalcophilic tendency. The element occurs most commonly along with sulfur in volcanic regions and readily forms lattices with sulfides (Kabata-pendias et al., 2001). Because of structural similarity between Se^{2-} and S^{2-} anions, selenium readily substitutes sulfur in the crystal structures of many sulfide minerals (Butterman et al., 2004).

In soils and sediments, selenium can be present in the soluble, exchangeable, organic matter-bound, sulfide, carbonate and oxide fractions of sequential extraction procedures. It occurs mainly in four stable oxidation states in the soil environment in the form of selenate (SeO_4^{2-},

13

Se(VI)), selenite (SeO_3^{2-}, Se(IV)), elemental selenium (Se(0)) and selenide (Se(-II)). Selenium oxyanions, selenate and selenite, are most bioavailable. SeO_4^{2-}, a stable form of Se in the soil and water environment (Yasin et al., 2014), is water soluble and thus readily absorbed by plant roots. SeO_3^{2-} is found adsorbed on organic matter in soil and forms strong stable complexes on the surfaces of minerals such as ferrihydrite, geothite, hematite, apatite and MnO_2 (Kabata-pendias et al., 2001). Thus, iron minerals influence the solubility of selenium oxyanions, particularly selenite in soils. For example, ferrihydrite ($Fe(OH)_2$) has a strong adsorption capacity for selenite and thus limits selenium mobility and bioavailability (Kausch et al., 2012).

Figure 2.3. Most common and naturally occurring selenium minerals in earth's crust viz. (a) klockmannite, (b) ferroselite, (c) naumannite, (d) clausthalite and (e) tiemannite. The crystal structure of each mineral is shown in the inset (Downs & Hall-Wallace, 2003).

The chemical forms of selenium and their solubility mainly depend on the prevailing redox conditions, pH, salinity and carbonate content. Other factors such as organic matter concentration and composition, iron concentration as well as clay type and content also contribute to absorption, bioavailability and toxicity (Bajaj et al., 2011; Schilling et al., 2015). In acidic clay soils and soils with a high organic matter content, selenium is mainly present in the form of selenides and selenium sulfides (SeS_2/Se_nS_{8-n}), which have limited solubility and bioavailability (Kabata-pendias et al., 2001). In well drained neutral soils, selenium is predominantly available in the form of selenite (Mayland et al., 1989). In alkaline and aerated soils, selenium occurs as selenate which is mobile and available to plants. Selenite is oxidised to selenate, the more mobile and bioavailable form of selenium in soil environments (Dhillon and Dhillon, 2014).

2.2.2. Sources of selenium in soil

Weathering of rocks is one of the major sources of environmental selenium (Wu 2004). Selenium rich rocks such as black shales, carbonaceous limestones, carbonaceous cherts, mudstones and seleniferous coals create selenium hotspots in the earth's crust (Winkel et al., 2012). Selenium is generally associated with the clay fraction of the soil and its concentration varies among different rocks with shales (600 μg Se g^{-1}), phosphatic rocks (300 μg Se g^{-1}) and sedimentary rocks containing higher amounts of selenium than igneous rocks or limestone and sandstone **(Table 2.1).**

Table 2.1. Representative selenium content in the earth's crust

Type of rock	Selenium content (mg L^{-1})			
	Butterman and Brown (2004)	Wu (2004)	Fordyce (2007)	Mayland et al. (1989)
Igneous rocks	0.05			
Sandstones	0 to 0.5	<0.1		0.05-0.08
Shales	0 to 0.6	0.6	600	0.6
Limestones	0.08	<0.1		0.03-0.10
Soils	0.2			
Phosphatic rocks			300	1-100
Black shales	20 - 1500			
Coals	0.5 to 12			1-20

In the western United States, the major source of selenium in soil has been the cretaceous shale deposits of the prehistoric inland sea (Hladun et al., 2013; Parkman & Hultberg, 2002). Emission of selenium into the atmosphere from natural deposits, anthropogenic activities, or volatilization from the soil and plants followed by atmospheric precipitation in the form of rain causes a major turnover in the selenium cycle in environment (Winkel et al., 2015). Recently, Sun et al. (2016) concluded that deposition and volatilization are key processes in regulating selenium concentrations in surface soils of China, rather than the nature of the parent rock material.

Selenium from the cretaceous shale deposits of the prehistoric inland sea dissolve in runoff from water-rock interactions and irrigation on selenium rich agricultural soils lead to enrichment of selenium levels in water and soil (Yasin et al., 2014). The contamination of surface water by selenium in the San Joaquin Valley of California (United States) and Punjab (India) are examples of selenium accession via water-rock interaction and irrigation with Se-rich groundwater, respectively (Winkel et al., 2012). Concentration of the most bioavailable form of selenium, i.e. water soluble selenate, may increase by phosphate fertilization with Se-containing phosphate rocks and may increase the selenium concentration in agricultural drainage (Wu, 2004) (**Figure 2.4**). In the USA, the major anthropogenic factors that lead to elevated selenium concentrations in soil up to toxic levels are combustion of fossil fuels and distribution of coal fly ash from thermal power stations on land as selenium leaches from these ashes (**Table 2.2**).

Table 2.2. Different selenium species and their important sources (Zhang & Moore, 1996)

Selenium Form	Sources
Selenate	Agricultural irrigation drainage
	Treated oil refinery effluent
	Mountaintop coal mining/ valley fill leachate
	Copper mining discharge
Selenite	Oil refinery effluent
	Fly ash disposal effluent
	Phosphate mining overburden leachate
Organoselenium (selenides)	Treated agricultural drainage (in ponds or lagoons)

2.2.3. Sources of selenium in sediments

Selenium in marine sediments is usually associated with effluent discharges from mining industries or coal fired power plants. Ellwood et al. (2016) studied the volatile selenium flux in one such selenium contaminated sediment with selenium concentrations of 3 to 6 µg Se g^{-1} in a coastal region of Lake Macquarie (NSW, Australia). Overflow from ash dams of the coal-fired power station into the lake lead to gradual methylation of selenium present in the sediments to volatile dimethyl selenide (DMSe). This natural remediation process led to the total loss of 1.3 kg Se per year by volatilization. About 100,000 cubic metre of high salt and Se-laden sediment from the San Luis Drain (California, USA), which had to be closed down due to adverse effects on waterfowl and fish in and around the Kesterson reservoir, is in need to be disposed safely. Selenium in the sediment ranges was as high as 57 ($\pm$ 6.1) µg Se g[-1]. A 7-year field study was carried out to understand the effect of disposing this Se-laden sediment to agricultural land by Bañuelos et al. (2013). Based on selenium measurements across soil, significant migration of selenium from sediment on the surface soil to lower depths in the soil was observed.

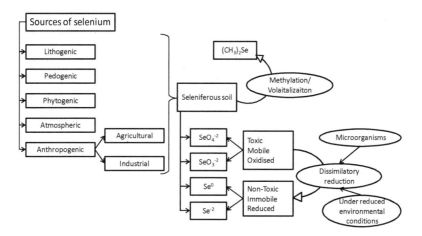

Figure 2.4. Illustration of various sources of selenium and Se transformations in the soil environment.

As an evidence of transport of selenium by aquatic pathways, a significant increase in selenium flux was also observed in sediment cores of lakes adjacent to the McClean lake uranium mines in north Saskatchewan (Canada) (Laird et al. 2014). The effect of selenium in the sediments of southwest England on the biota, i.e. macroalgae and invertebrates, has been investigated by Turner (2013). The selenium concentration in macroalgae (0.05 to 0.51 µg Se g[-1]) was found to be lower than in the sediment (0.4 to 1.49 µg Se g[-1]), while that in invertebrates (2.82 to 12.68 µg Se g[-1]) was found to be significantly higher than in the sediment or algae, suggesting trophic transfer and bioaccumulation of selenium in higher trophic levels. Selenium speciation and concentrations in San Francisco bay sediments (0.16 to 0.98 µg Se g[-1]) had a positive correlation between the total selenium and organic carbon concentration in marine sediments, probably due to incorporation of selenium as selenocysteine in proteins in marine phytoplankton proteins (Meseck and Cutter, 2012). A significant flux of selenium from the sediment to marine water was also recorded.

2.3. The biogeochemical selenium cycle

2.3.1. Selenium speciation

Selenium distribution and speciation depends on a variety of physical, chemical and biological processes, such as volcanic activity, combustion of fossil fuels, weathering of rocks and soils, soil leaching, groundwater transport, plant and animal uptake and release, adsorption and desorption, chemical or biological reduction and oxidation as well as mineral formation (Wu, 2004). Selenium exposure and toxicity to animals primarily results from selenium biotransformation and transfer pathways through the food web. The aim of selenium remediation is not only to remove excess selenium from the soil environment, but also to reduce ecotoxicological risk. Bioavailability of selenium is governed by selenium species present in soil and therefore factors that affect the selenium dynamics (i.e. mobility, solubilisation and transformations) in the soil should be considered. Monitoring of the total selenium concentration alone in soil is, therefore, not a reliable parameter for success of selenium bioremediation technologies (Frankenberger et al., 2004). It is important to study transfer of selenium from the soil to plants and higher trophic levels, selenium migration in soil or volatilization as well as selenium concentrations in representative tissues and organs in higher animals (Ashworth and Shaw, 2006).

Recently, the selenium distribution and translocation from soil to the barley plant in the Tibetan Plateau Kashin-Beck disease (KBD) area was compared with that of a non-KBD area (Wang et al., 2016). KBD is a chronic, endemic osteochondropathy (bone disease) associated with selenium deficiency (see below). Although the soil selenium content in both the KBD and non-KBD area was almost similar, uptake of selenium by crop plants was significantly lower in the KBD area as compared to the non-KBD area. This was mainly attributed to the difference in soluble, exchangeable and fulvic acid bound selenium. The authors suggested that this can be rectified by improving selenium-bioavailability by increasing soil pH and reducing the organic matter content. In a long term (>20 years) study of selenium accumulation in soil and uptake by plants, Wang et al. (2016) also showed that an increase in organic carbon content of soil led to a decrease in selenium absorption by plants. However, the exchangeable selenium content of soil can be increased by increasing potassium levels in the soil. Therefore, it is essential to consider selenium speciation in addition to concentration while understanding selenium toxicity on biota.

2.3.2. Isotopic speciation of selenium

Six major isotopes of selenium are found in nature. These are, along with their weight abundances, ^{74}Se (0.89%); ^{76}Se (9.36%); ^{77}Se (7.63%); ^{78}Se (23.78%); ^{80}Se (49.61%); and ^{82}Se (8.73%). Out of these, ^{82}Se is a radioactive isotope with a half-life of 1.08×10^{20} years (Butterman et al., 2004). Selenium isotopes in natural environments have been widely studied as an indicator of selenium migration in the soil-water environment. The effect of biogeochemical cycling of selenium in a selenium-contaminated lake and littoral wetland at Sweitzer lake (Colorado, USA) was studied by tracing selenium stable isotopes. The study revealed an interesting finding that selenite in the lake originates primarily from the oxidation of organically bound selenium rather than the reduction of selenate (Clark and Johnson, 2010).

Selenium isotopes have also been used to study the accumulation of selenium in the wetland sediments at San Jaoquin valley (California, USA). Selenium was found to be assimilated in plants and algae followed by its deposition and mineralisation (Herbel et al., 2002). Migration and immobilisation of selenium isotopes at different redox states has been investigated in seleniferous soil in Punjab (India) by Schilling et al. (2015). Migration of isotopically lighter selenium (organically bound Se) downward through soil profile was significantly higher than that of isotopically heavier selenium (water-soluble Se). A positive correlation between the carbon to nitrogen ratio and organically-bound selenium isotopes was also observed. Not only does the selenium concentration but also the redox state of selenium play an important role to determine the effect of selenium on bio-absorption and biomagnification. This analysis, therefore, provided an insight into the biogeochemical processes of agricultural seleniferous soils and could benefit for assessment of better soil management with regards to selenium soil concentrations and redox states.

2.3.3. Factors affecting selenium speciation in soil

Soil pH and redox potential (**Figure 2.5**), fixation capacity, competing ions (mainly phosphates and sulfates), clay type, organic matter and microbiological activity may affect speciation and thus, mobility of selenium in soil (**Table 2.3**). The total selenium content varies with the organic matter and clay content along the soil profile. Selenium binds to proteins, fulvic acids, clay and organic matter in acidic soils (Cuvardic 2003; Wu 2004).

Table 2.3. Soil conditions that affect the speciation, mobility and bioavailability of selenium (Kabata-pendias et al., 2001)

Soil factor	Variables	Major Se form	Mobility
pH	Alkaline	Selenate	High
	Neutral	Selenite	Moderate
	Acidic	Selenides	Low
Redox potential	High	Selenite	High
	Low	Selenides	Low
Organic matter	Undecayed	Adsorbed	Low
	Decayed	Complexed	High
	Enhanced biomethylation	Volatilized	High
Clay content	High	Adsorbed	Low
	Low	Soluble	High

Selenium biogeochemical reactions are balanced by redox reactions, where its solubility increases under high redox potential. The mobility of selenium with respect to adsorption is favoured by alkaline pH, oxidising agents and high concentration of additional anions. Adsorption of selenium onto the soil minerals (oxides, hydroxides, or clays) decreases with an increase in pH. Anions such as CO_3^{2-}, Cl^-, SO_4^{2-} and PO_4^{3-} compete with Se(IV) and Se(VI) for adsorption sites and, therefore, decrease selenium sorption by direct competition (Goh et al., 2004). In oxic surface soils, selenite (Se(IV)) and selenate (Se(VI)) are the major selenium species. Their reduction into solid, insoluble Se(0) and metal-Se(−II) precipitates (e.g., FeSe, $FeSe_2$) can occur in local anoxic zones (e.g., soil aggregates) due to either low redox potential values or reaction with reduced mineral phases (Tolu et al., 2014).

Selenium cycling in the environment depends on its speciation, redox states and its bioavailability which in turn depends on its organic or inorganic nature. In order to quantify selenium abundance in operationally defined fractions, several soil scientists have developed various sequential extraction protocols that utilize reagent solutions to separate the different forms of sedimentary selenium on the basis of solubility, exchangeability, volatility, and redox activity (Hagarova et al., 2003; Kulp and Pratt, 2004; Martens and Suarez, 1997). A summary of the methods employed for sequential extraction of selenium species from soil is given in (**Table 2.4**).

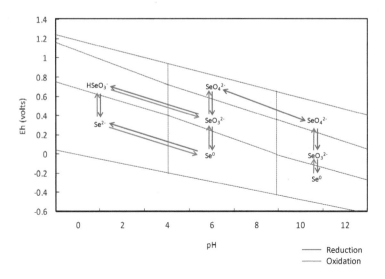

Figure 2.5. Effect of pH and Eh of soil on selenium speciation (Mayland et al., 1989)

Table 2.4. Summary of sequential extraction procedures used for characterization of selenium species in different soils

Soil Characteristics and Geographic Area	Washing Steps	Selenium in different forms separated	References
Upper Cretaceous chalk and shale, South Dakota and Wyoming, USA	6	Soluble Se(IV), Se(VI); ligand-exchangeable, tightly held Se(IV); base-soluble inorganic and organic selenides; organic Se or selenide minerals; selenate ions substituted in the carbonate atomic lattice in Calcite crystals; ferroselite (FeSe2); Se substituted for S in pyrite	Kulp and Pratt (2004)
Evaporation drainage pond soils contaminated with Se-laden drainage water, San Joaquin Valley of California	4	soluble Se; adsorbed Se; "organically associated" Se; "refractory" Se (including Se0)	Martens and Suarcz (1997)
Wetland system containing moderate levels of selenium, Benton Lake National Wildlife Refuge (Montana)	5	Soluble; adsorbed; elemental; organic material; easily reducible, crystalline and amorphous oxides	Zhang and Moore (1996)

Synthetic soil spiked with Se species Core and "grab samples" from constructed wetlands site, Tulare Lake Drainage District, Corcoran CA, USA Stewart Lake Utah, USA Sediment samples (0-30 cm depth) from two dewatered ponds, Kesterson National Wildlife Refuge Los Banos, CA, USA Sediments, Benton Lake Montana, USA	5	"Soluble/exchangeable" Se; "Adsorbed" Se; "Organically associated" Se; "Elemental" Se; "More recalcitrant forms" of organic Se	Wright et al. (2003)
Selenium contaminated sediments, Stewart Lake Wetland, California	4	"Soluble" Se; "Adsorbed" Se;"Organically associated" Se; "Elemental" Se	Ponce de Leon et al. (2003)
Calcaro-haplic chernozem; pH 5.93, Trnava, Slovakia Stagno-gleyiv luvisol; pH 6.54, Zvolen, Slovakia	6	Ion exchangeable Se and bound to carbonate; Se in reductive phase bound to Mn-Fe oxides; Se bound to organic matter; Se bound to humic compounds; Se bound to sulfides.	Hagarova et al. (2003)

2.3.4. Role of the soil compartment in the biogeochemical selenium cycle

The amount of selenium in the food chain, and thus in the human diet, depends on the selenium concentrations in the soil. Therefore, soil is the most important part of the environment in selenium cycling (Hagarova et al., 2005), although the selenium cycle occurs across the lithosphere, hydrosphere and atmosphere. Se(IV) is considered more toxic to the microorganisms than Se(VI) because it is easily adsorbed onto organic matter and inorganic minerals present in the environment making it more bioavailable. The insoluble forms of selenium, elemental selenium (Se(0)) and selenide (Se(−II)) are formed by the reduction of selenium oxyanions. The reduction can occur chemically in reducing environmental conditions or by microbial assimilatory and dissimilatory reduction mechanisms (see below). Application of selenate and selenite reduction by microorganisms for bioremediation of seleniferous soil and water in natural environments is being widely studied (Hagarova et al., 2005; Kausch et al. 2012).

Table 2.5. Selenoproteins and their functions in prokaryotes

Selenoprotein	Function	Organisms	References
Formate dehydrogenase	Oxidation of formate to carbon dioxide with molybdenum as co-factor	*Methanococcus jannaschii, Moorella thermoacetica*	Stock and Rother (2009)
Glycine reductase (GrdA & GrdB)	GrdA & GrdB for two subunits of glycine reductase enzyme are substrate specific	*Clostridium difficile, Clostridium sticklandii*	El-Aassar et al. (2008), Stock and Rother (2009)
Proline reductase	Membrane-bound, Sec-containing enzyme, reduction of D-proline to 5-animovalerate	*Clostridium difficile,, Clostridium sticklandii*	Kryukov and Gladyshev (2004), Stock and Rother (2009)
Selenophosphate synthetase	Function unknown	*Syntrophus aciditrophicus, Desulfotalea psychrophila*	Zhang et al. (2006)
Formylmethanofuran dehydrogenase	Oxidation of formymethanofuran with molybdenum in enzyme active site	*Methanopyrus kandleri*	Lenz (2008)
Heterodisulfide reductase	Membrane-bound, cytochrome-be containing Sec-independent enzyme involved in electron transport.	*Desulfotalea psychrophila. Methanococcus jannaschii*	Foster (2005), Zhang et al. (2006)
Thioredoxin (Trx)	Major intracellular protein disulfide reductant	*Treponema denticola*	Kryukov and Gladyshev (2004), Stock and Rother (2009)
Sarcosine reductase	Deanination of sarcosine to monomethylamine	*Eubacterium acidaminophilum*	Stock and Rother (2009)
Betaine reductase	Deamination of betaine to acetyl-phosphate and trimethylamine	*Eubacterium acidaminophilum*	Stock and Rother (2009)

Xanthine dehydrogenase	Molybdo-flavoenxyme. Deamination and decarboxylation of xanthine	*Clostridium acidiurici*	Stock and Rother (2009)
Purine hydroxylase	Molybdo-flavoenxyme. Deamination and decarboxylation of xanthine	*Clostridium purinolyticum*	Stock and Rother (2009)
Nicotinic acid dehydrogenase	Degradation of nicotinate to pyruvate and propione	*Methanosarcina barkeri*	Stock and Rother (2009)
Peroxiredoxin (Prx)	Abundant Cys-containing enzyme. Function unknown.	*Eubacterium acidaminophilum, Syntrophobacter fumaroxidans*	Foster (2005), Zhang et al. (2006)
Glutathione peroxidase	Function unknown	*Treponema denticola*	Kryukov and Gladyshev (2004), Stock and Rother (2009)
Glutaredoxin (Grx)	Disulfide oxidoreductase. Functions independent of glutathione	*Geobacter metallireducens*	Foster (2005),Kryukov and Gladyshev (2004), Stock and Rother (2009)
HesB-like	Iron-sulfur cluster biosynthesis	*Methanococcus jannaschii, Syntrophobacter fumaroxidans*	Stock and Rother (2009)

The soluble species of selenium (SeO_4^{2-} and SeO_3^{2-}) are bioavailable to the microbial community in soil. Assimilatory reduction pathways are used for the reduction of selenium oxyanions, to synthesize SeCys or to form selenoproteins (**Table 2.5**) in microorganisms (Nancharaiah and Lens, 2015a; Lenz et al., 2009). Soluble selenium oxyanions are converted to insoluble selenium forms by different mechanisms. Microorganisms have developed mechanisms to decrease the toxicity by dissimilatory reduction and other modes of detoxification of the soluble selenium oxyanions to insoluble, relatively non-toxic elemental selenium (Hagarova et al., 2005). Chemical reduction of SeO_4^{2-} and SeO_3^{2-} has been observed in acidic soils with lower redox potential. However, Se(VI) does not readily undergo chemical reduction under ambient pH and temperature conditions (Stolz et al., 1999). The soluble forms of selenium have a tendency to contaminate water bodies and leach out from the topsoil into groundwater by percolation due to rainfall or irrigation of agricultural lands (**Figure 2.6**).

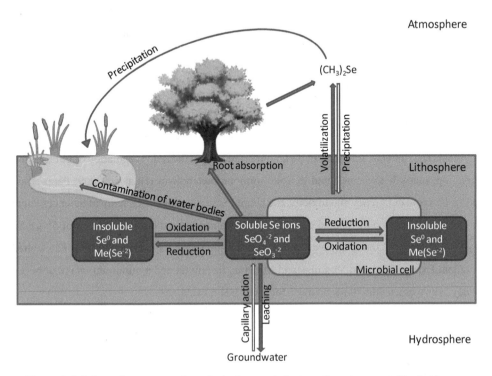

Figure 2.6. Schematic representation of selenium cycle in the soil environment. The bold arrows indicate preferred direction of the process (Adapted from Flury, Jr, and Jury 1997)

One of the major processes in the biogeochemical cycle of selenium particularly in soils and rhizosphere environments is selenium biomethylation, which forms volatile selenium compounds (Winkel et al., 2010). Selenium undergoes microbial transformations or biomethylation to form volatile compounds such as DMSe, dimethyl selenenyl sulphide (DMSeS) or dimethyldiselenide (DMDSe) (Flury et al., 1997; Dungan et al., 2003; Winkel et al., 2010). Biological oxidation of elemental selenium to selenate and selenite occurs under aerobic conditions by selenium oxidising microorganisms (Dowdle and Oremland, 1998). Chemical oxidation of the reduced forms of selenium occurs naturally in the soils with alkaline pH and high redox potential, while chemical reduction of selenium oxyanions occurs in acidic soils with low redox potential (Moreno et al., 2013).

Selenium in soil has different origins such as lithogenic, pedogenic, atmospheric, phytogenic and anthropogenic (Kabata-pendias et al., 2001). It was estimated that about 37-40% of the total selenium emissions to the atmosphere are due to anthropogenic activities such as selenium applied to soils as foliar sprays, seed treatments, and phosphate fertilizers for agriculture as well as release of industrial effluents, dumping of fly-ash, smelting waste and wastes of ores (El-Ramady et al., 2015).

2.4. Selenium essentiality

2.4.1. Metabolic role of selenium in animals and humans

In animals and humans, selenium plays an important role in the redox regulation of intracellular signalling, redox homeostasis and thyroid hormone metabolism (Huawei, 2009; Papp et al., 2007). To avoid deficiency and toxicity, the United States National Academy of Sciences Panel on Dietary Oxidants and Related compounds recommended a dietary allowance of 55 µg Se day^{-1} in humans and set an upper tolerable limit of 400 µg Se day^{-1} (WHO 2011). The United Kingdom Expert Group on Vitamins and Minerals recommended a minimum intake of 60 µg Se day^{-1} for women and 70 µg Se day^{-1} for men (UK EVGM, 2003).

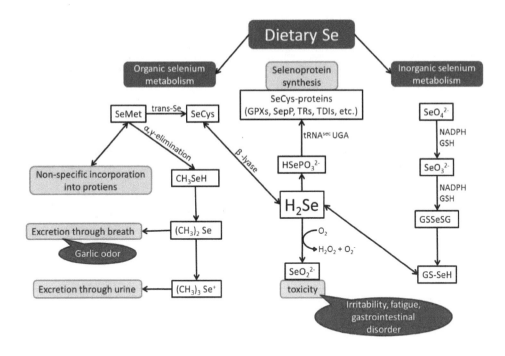

Figure 2.7. Schematic representation of selenium metabolism in mammals. Se, selenium; SeMet, selenomethionine; SeCys, selenocysteine; H_2Se, hydrogen selenide; $HSePO_3^{2-}$, selenophosphate; CH_3SeH, methylselenol; $(CH_3)_2Se$, dimethyl selenide; SeO_2, selenium dioxide; $(CH_3)_3Se^+$, trimethyl selenonium ion (Tinggi, 2003; Fairweather-tait et al., 2010; Huawei, 2009)

Table 2.6. List of selenoproteins identified in humans and their functions

Selenoprotein	Tissue/position	Role/ Functions	References
Glutathione peroxidase (GPx1)	Cell cytosol	Antioxidant reducing H_2O_2 and phospholipase A2 cleaved lipid hydroperoxides and storage vehicle for Se	Brown & Arthur (2001); Diamond (2015); Papp et al. (2007); Rayman (2012)
Glutahione peroxidase (GPx2)	Gastrointestine	Protection from toxicity of ingested lipid hydroperoxides in mammals; intracellular defence mechanisms against oxidative damage by preventing production of reactive oxygen species in colon	Brown & Arthur (2001); Papp et al. (2007); Rayman (2012)

Glutathione peroxidase (GPx4)	Membrane associated phospholipid hydroperoxide.	Reductive destruction of lipid hydroperoxides and small soluble hydroperoxides; capable of metabolising cholesterol and cholesterol ester hydroperoxides in oxidised low density lipoprotein	Brown & Arthur (2001); Diamond (2015); Papp et al. (2007); Rayman (2012)
Glutathione peroxidase (GPX3)	Extracellular	Possible antioxidant in renal tubules	Brown & Arthur (2001); Papp et al. (2007); Rayman (2012)
Selenium-binding protein 1 (SBP1)	Nucleus and cytoplasm of prostrate tissue	Toxification/detoxification process, cell-growth regulation, intra-Golgi protein transport, aging and lipid metabolisms	Diamond (2015); Papp et al. (2007);
Thioredoxin reductases-1 (TR1)	Cell cytosol	Regulation of intracellular redox state	Papp et al. (2007); Rayman (2012)
Thioredoxin reductases-2 (TR2)	Mitochondria	Regulation of intracellular redox state	Papp et al. (2007); Rayman (2012)
Thioredoxin reductases-3 (TR3)	Testis	Regulation of intracellular redox state	Papp et al. (2007); Rayman (2012)
Iodothyronine deiodinases type1 (DI1)	Kidney, liver, thyroid, brown adipose tissue	Inactive thyroxine metabolism to active 3,3'-5'triiodothyronine	Papp et al. (2007); Rayman (2012)
Iodothyronine deiodinases type2 (DI2)	Thyroid, central nervous system, pituitary, skeletal muscle, adipose tissue	Activation of thyroid hormones	Papp et al. (2007); Rayman (2012)
Iodothyronine deiodinases type3 (DI3)	Placenta, central nervous system, fetus	Inactivation of thyroid hormone	Papp et al. (2007); Rayman (2012)
Selenoprotein-P (SelP)	Plasma, brain, liver and testis	Selenium homeostasis; antioxidant activity	Brown & Arthur (2001); Papp et al. (2007); Rayman (2012)

Selenoprotein-W (SelW)	Brain, colon, heart, skeletal muscle, prostrate	Antioxidant activity, cardiac and skeletal muscle metabolism	Brown & Arthur (2001); Papp et al. (2007);
Selenoprotein-N (SelN)	Endoplasmic reticulum	Unknown	Papp et al. (2007)
Selenoprotein-S (SelS)	Endoplasmic reticulum	Role in innate immune response, inflammation, regulation of cytokines	Papp et al. (2007)
Selenoprotein-K (SelK)	Endoplasmic reticulum	Unknown	Papp et al. (2007)
Selenoprotein-R (SelR)	Cytosol and nucleus	Methionine metabolism, protein repair, antioxidant activity	Papp et al. (2007)
Selenoprotein-H (SelH)	Nucleus	DNA binding protein, regulation of glutathione synthesis genes and phase II detoxification.	Kurokawa & Berry, (2013); Fairweather-tait et al., (2010)
Selenoprotein-M (SelM)	Moderate expression on heart, lungs, kidney, uterus and placenta; and high expression in thyroid and brain	Unknown	Papp et al. (2007)
Seleno-phosphate synthetase	Cell cytosol	This enzyme regulates selenocysteine incorporation in selenoproteins to prevent toxicity	Papp et al. (2007)

In humans, selenium plays a complex metabolic role in protection of body tissues against oxidative stress, maintenance of the immune system and modulation of growth and development (**Figure 2.7**). Most of the assimilated selenium in tissues is available in the form of proteins called selenoproteins, in which selenium exists as selenocysteine (SeCys). **Table 2.6** lists selenoproteins identified in humans and their functions. Selenium is an essential part in the antioxidant enzyme glutathione peroxidase which protects cell membranes from damage caused by lipid peroxidation. Selenoproteins like thioredoxin reductase and glutathione peroxidase play a role in cancer prevention by preventing intracellular oxidative stress (Kalender et al., 2013; Selenius et al., 2010). Selenoproteins play an important role not only in the regulation of the intracellular redox state and anti-inflammatory functions, but may also play a significant role in glucose metabolism, calcium metabolism and glycoprotein folding (Rayman, 2012). Selenium forms a structural component of specific selenoproteins incorporated in the form of Se-methionine (Se-Met) in plants and Se-cysteine (Se-Cys) in animals. Active sites of selenoproteins consisting of SeCys have redox functions, such as scavenging free radicals (**Figure 2.8**), thus preventing oxidative stress and cancer (Misra et al., 2015; Tapiero et al., 2003).

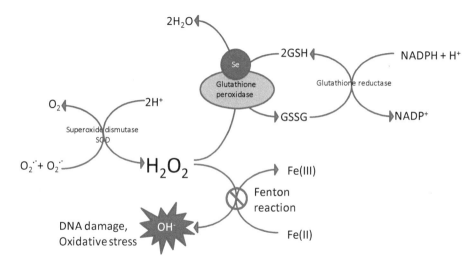

Figure 2.8. Role of selenoprotein glutathione peroxidase in scavenging free radicals (Navarro-Alarcon and Cabrera-Vique, 2008)

2.4.2. Selenium toxicity

Selenium is a contaminant of potential environmental concern and is hence one of the most important elements from environmental pollution point of view. Formerly, the element was considered to be toxic when the livestock grazing on grass grown in selenium-enriched soil developed disorders known as alkali disease and blind staggers in 1930 in South Dakota (Tinggi, 2003). Selenium has a tendency to bioaccumulate in the aquatic environment and becomes toxic to aquatic organisms such as fish at elevated concentrations. Selenium becomes toxic to cormorants and other birds that prey on aquatic organisms harbouring elevated levels of this element (Miller et al., 2013). In 1984, high incidences of deformities and mortalities were recorded in water fowls in the Kesterson wildlife reservoir (California, USA), where agricultural drainage water and industrial effluent containing a high selenium content was discharged.

Selenium bioaccumulation has exposed the fish and water fowl in wetlands and evaporation ponds to a severe threat, which caused deformities, impaired reproduction and eventually death of fish and birds (Zhang et al., 2008). Transfer of selenium to the higher tropic levels and accumulation in the food chain due to selenium bioaccumulation (Riva et al., 2014) caused additional damaging effects such as nausea, vomiting and diarrhea due to selenium poisoning (selenosis) in humans and animals along with dermal and neurological dysfunction associated with deformation of nails and hoofs, unsteady gait or even paralysis and cardiovascular symptoms (Duntas and Benvenga, 2014). Dietary uptake higher than 400 µg day^{-1} was linked to hair and nail loss and disruption of the nervous and digestive systems in humans and animals (Misra et al., 2015). Lemly (2014) calculated the monetary loss due to reduced fish productivity and studied the impact of selenium pollution in water bodies associated with high selenium concentrations on tissues of aquatic organisms (e.g. fish), particularly morphological abnormalities and teratogenic deformities.

2.4.3. Selenium deficiency

The perspective of the researchers towards selenium changed in 1960, with the identification of a peculiar heart muscle disease symptom called Keshan's disease in selenium deficient populations in China (Chen, 2012). A selenium concentration in staple food lower than the critical standard of 100 µg kg^{-1} can lead to development of selenium deficiency diseases (Wang et al., 2016). Selenium deficiency causes reproductive disorders and heart failure in humans, white muscle disease in young animals, fatal diseases such as hepatosis and Mulberry heart disease in pigs and exudative diathesis in poultry (Mehdi et al., 2013). High incidences of white muscle disease and heart necrosis were observed in selenium deficient sheep and cattle in New Zealand and Western Oregon, USA (Tinggi, 2008) . These and other studies have contributed to the understanding of physiological functions of selenium in higher animals and humans. The narrow window of 40 – 400 µg day^{-1} between selenium deficiency and toxicity has led to selenium be appropriately termed as an 'essential toxin' (Lenz and Lens 2009) and makes environmental selenium research (Winkel et al., 2012) critical to maintain a balance between providing the necessary and to avoid potential toxicity.

2.5. Plant - selenium interactions

2.5.1. Selenium metabolism in plants

In plants, selenium is metabolized by mechanisms similar to those of sulfur metabolism (**Figure 2.9**). The uptake and translocation of selenate in plants is significantly faster than that of selenite (Kaur et al., 2014; Yu & Gu, 2008). Selenate is taken up by the plants via high affinity sulfate transporters in the roots and assimilated into SeCys and selenomethionine (SeMet) (Li et al., 2008; Pilon-Smits et al., 2009). Selenium has a beneficial role in plant growth as well as pathogen and herbivore resistance in selenium hyperaccumulator plants (Yasin et al., 2014). Saidi et al. (2014) investigated the beneficial effect of selenium by alleviating cadmium toxicity by preventing oxidative stress in sunflower (*Helianthus annuus*) seedlings. Hasanuzzaman et al. (2010) illustrated the regulatory role of selenium in various physiological processes in plants as well as its protective role under abiotic stress conditions. Compared to non-accumulators, selenium hyperaccumulator plants (genera *Stanleya* and *Astragalus*) are capable of accumulating significantly higher amounts of selenium which may account upto 1% of their dry weight (Tamaoki et al., 2008).

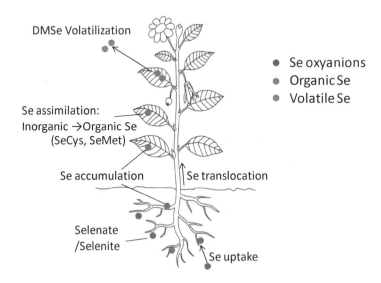

Figure 2.9. Movement and possible fate of selenium in plants (Adapted from Pilon-Smits et al., 2009)

Selenium also plays a positive role in enhancing the antioxidant activity through selenoproteins in soyabean (Hu et al., 2014). In this study, selenoproteins isolated from soyabean growing in seleniferous soil consisted of 18 times higher selenium content than control plants. The radical scavenging activity of these selenoproteins was also found to be significantly higher (about 4-fold) than that of control proteins. The impact of seleniferous soils on interactions between plant and pollinators in the moderate selenium accumulating *Raphanus sativus* L. (radish plant) was evaluated by Hladun et al. (2013) in a greenhouse study. Application of selenium on this cross-pollinating plant species caused an increase in seed abortion and decrease in plant biomass and herbivory by birds and aphids. While an increase in plant biomass and a decrease in herbivory was observed in both *Astragalus* hyperaccumulator and non-accumulator plants (Statwick et al., 2016).

In food crops, selenium when used at optimum doses benefits different inter-related plant physiological processes. Selenium exhibits anti-oxidant activity protecting potatoes against abiotic stress such as cold and light (Seppänen et al., 2003), sorghum species against heat stress (Djanaguiraman et al., 2010) and *Acer saccharium* against desiccation (Pukacka et al., 2011). The decrease in reactive oxygen species under low selenium dose enhances photosynthesis, thereby promoting growth in sorghum and rice (Wang et al., 2012). Selenium also assists in delaying senescence in rye grass, lettuce, soybean (Xue et al., 2001), while increasing the carboxylation efficiency in tobacco further promoting plant growth (Jiang et al., 2015).

High concentrations of selenium also lead to selenium toxicity in crop plants (Leduc et al., 2004). Reduction in brown mustard, corn, rice and wheat crop yields was demonstrated by Rani et al. (Rani et al., 2005) due to a high selenium content in the soil. Especially wheat plants showed visible symptoms of selenium toxicity such as snow white chlorosis with rose colouration on leaf sheath (Dhillon and Dhillon, 1991). In plants, selenium toxicity occurs either due to deformation of selenoproteins or oxidative stress (Gupta and Gupta, 2017). Chatelian et al. (Châtelain et al., 2013) observed that replacement of cysteine with SeCys in the methionine sulfoxide reductase enzyme leads to its malformation and hinders enzyme function in *Medicago truncatula* and *Arabidopsis thaliana* plant seeds. As compared to cysteine, SeCys is larger, more reactive and easily deprotonated. SeCys substitution distorts the tertiary structure of the protein due to diselenide bridge formation and SeCys substitution affected the enzyme kinetics as well (Hondal et al., 2013). At higher concentrations, Se acts as a pro-oxidant and generates reactive oxygen species in plants (Gomes-Junior et al., 2007).

Increased lipid peroxidation and cell mortality was observed in *Arabidopsis* under selenium stress (Lehotai et al., 2012).

2.5.2. Role of rhizospheric microorganisms

Bacteria found in the rhizospheric region of selenium hyperaccumulative plants can enhance selenium reduction during phytoremediation. The selenium accumulating capacity of the crop plant *Brassica juncea* was studied by inoculating with a Se-tolerant bacterial consortium. Plant growth promoting properties such as improving availability of mineral nutrients to plants, increasing biological nitrogen fixation and production of plant growth regulators were enhanced by addition of the consortium (Yasin et al., 2014). Rhizospheric, nitrogen-fixing bacteria such as *Azospirullum brasilense* (Tugarova et al., 2014), and *Rhizobium selenireducens* (Hunter et al., 2007) were able to reduce selenite and form selenium nanoparticles. A bacterial endophyte, *Pseudomonas moraviensis*, isolated from the selenium hyperaccumulator plant *Stanleya pinnata* was found to tolerate more than 100 mM of selenate and selenite and is capable of reduction of 10 mM selenite to elemental selenium in 48 h (Staicu et al., 2015).

Selenium hyperaccumulators naturally harbour an endophytic bacterial community in the rhizosphere that is significantly more diverse to that of the non-accumulators. The endophytic bacterium not only renders the hyperaccumulator with characteristics such as high selenium resistance, but also capacity to produce elemental Se and plant growth promoting properties (Sura-de Jong et al., 2015). The selenium tolerant bacterial strain *Microbacterium ozydans* was isolated from the rhizosphere of hyperaccumulator *Cardamine hupingshanesis* which accumulates selenium in the form of selenocystine ($SeCys_2$) (Tong et al., 2014). This strain was able to tolerate 15 mg L^{-1} selenite and a significant amount of $SeCys_2$ production was observed in the medium when exposed to 1.5 mg L^{-1} selenite. This endophytic bacterial strain is likely to play a prominent role in $SeCys_2$ accumulation in the plants.

2.6. Microbe - selenium interactions

2.6.1. Selenium metabolism in microorganisms

Selenium biotransformations in the environment are mainly carried out by selenium metabolizing microbes which can be categorized as selenium reducing and selenium oxidising bacteria (**Figure 2.10**). Selenium reduction from toxic selenium oxyanions to the less toxic elemental selenium form is a favoured reaction in anoxic environmental settings. Certain microorganisms have the ability to incorporate selenium into organo-Se(−II) compounds such as seleno-cysteine via the mechanism known as assimilatory reduction (Lenz and Lens, 2009; Sarret et al., 2005). Bacteria capable of dissimilatory selenate reduction are abundant in nature and have a constitutive ability to reduce selenate (Stolz et al., 1999). However, at high selenium concentrations, microorganisms reduce selenium oxyanions under both aerobic and anaerobic conditions as part of detoxification strategies (Hagarova et al., 2005).

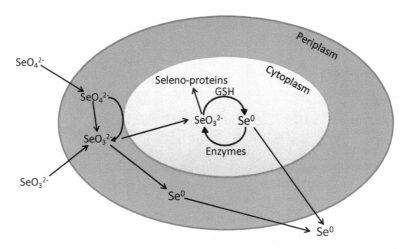

Figure 2.10. Selenium metabolism pathway and schematic representation of selenium reduction reactions in microorganisms (Adapted from Nancharaiah & Lens, 2015a)

Amorphous Se(0) nanoparticles formed from microbial reduction of selenium oxyanions show colloidal properties (Staicu et al., 2014). The reduction is due to membrane bound reductase enzymes causing the reduction either in the periplasmic space or outside the cell (Losi and Frankenberger Jr., 1997). Dissimilatory reduction of selenite is also mediated through this system (D.-B. Li et al., 2014). During this reduction process, extracellular formation of Se(0) nanoparticles is favoured over intracellular accumulation (Oremland et al., 2004). The extracellular formation of Se(0) nanoparticles is associated directly to the electron transport pathway and anaerobic respiration, while the intracellular accumulation is majorly due to the thiol mediated detoxification of selenite (Oremland et al., 2004).

Different hypotheses have been put forth for the release of Se(0) nanoparticles from microorganisms. One is the formation of selenium containing protrusions in *E. cloacae* SLD1a-1 where Se(0) nanospheres formed anaerobically were released via a rapid expulsion process

(Losi et al., 1997). Extracellular release of Se(0) nanoparticles observed by *Rhodospirullum rubrum* is proposed via a vesicular secretion system (Kessi and Hanselmann, 2004), while the release of Se(0) nanoparticles into media containing *D. sulfuricans* is due to cell lysis. However, these hypotheses require support due to lack of substantial experimental evidences (Tomei et al., 1995). Hence, further elucidation over establishment of the formation, transport and excretion mechanisms is required.

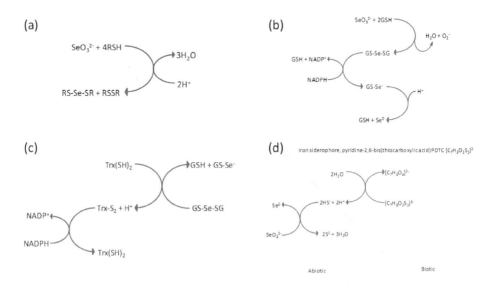

Figure 2.11. Selenium metabolic reactions observed in microorganisms: (a) Painter-type reaction, (b) Glutathione reductase reaction, (c) Thioredoxin reductase system and (d) Siderophore-mediated reduction

Several studies have demonstrated reduction of selenium oxyanions by microorganisms through detoxification mechanisms (Bajaj and Winter 2014; Ikram and Faisal 2010; Hunter and Manter 2008). Selenite detoxification by microorganisms has been proposed to occur by multiple mechanisms such as the Painter type reaction, modified Painter and Ganther reaction with glutathione reductase (Kessi and Hanselmann 2004), thioredoxin reducatase mediated reduction and iron-siderophore mediated reduction (Nancharaiah and Lens 2015a) (**Figure 2.11**). Garbisu et al. (1999) have proposed a quorum-sensing mechanism to detoxify selenite involving a dithiol system in *B. subtilis*. In response to high levels of selenite in the medium, an increase in cellular concentrations of thioredoxin and NADP-thioredoxin reductase was observed in *B. subtilis* along with morphological changes.

2.6.2. Aerobic reduction

Several studies on selenite reduction by microorganisms under aerobic conditions have been carried out (**Table 2.7**). Aerobic reduction of selenite and more than 24 h delay in appearance of selenium nanoparticles (SeNPs) by a strain of *Bacillus sp.* SeITE01, isolated from the rhizosphere of the Se-hyperaccumulator legume plant *Astragalus bisulcatus*, was reported (Lampis et al. 2014). The delay in intracellular and extracellular formation of SeNPs was attributed to the formation of an intermediate selenium compound. Aerobic reduction of selenite to intracellular SeNPs (50 to 400 nm in diameter) by two strains of the rhizosphere microorganism *Azospirullum brasilense* was described (Tugarova et al., 2014). Aerobic reduction of selenite to elemental selenium by *Pseudomonas alcaliphila* followed by the structural characterization of the bioreduced selenium showed formation of Se(0) nanorods with 50 to 500 nm diameter and stable spherical Se(0) nanoparticles with a diameter of about 200 nm after 24 h incubation (Zhang et al., 2011).

Table 2.7. Selenite reducing microorganisms isolated and characterized from seleniferous soils or selenium hyper accumulating plants

Organism	Origin	Reference
Selenite respiring bacteria		
Pseudomonas sp. strain RB	soil (used for illegal dumping of electronic appliances)	Ayano et al. (2014)
Azospirullum brasilense	rhizosphere	Tugarova et al. (2014)
Paenibacillus selenitireducens	selenium mineral soil	Yao et al. (2014)
Selenite detoxifying bacteria		
Bacillus mycoides SeITE01	rhizosphere of the Se-hyperaccumulator legume *Astragalus bisulcatus*	Lampis et al. (2014)
Bacillus megaterium	mangrove soil	Mishra et al. (2011)
Pseudomonas sp.	soil	Hunter and Manter (2009)
Rhizobium selenireducens sp.	rhizosphere	Hunter et al. (2007)
Stenotrophomonas maltophilia SeITE02	selenium- contaminated mining soil in Italy	Antonioli et al. (2007)

Gram positive, moderately halotolerant *B. megaterium* strains, isolated from mangrove soils showed resistance and potential to reduce selenite to elemental selenium under aerobic conditions (Mishra et al., 2011). *Pseudomonas* sp. Strain CA5 isolated from soil can withstand high selenite concentrations under aerobic conditions and reduce up to 66% of selenite from medium containing 150 mM selenite (Hunter et al., 2009). Interestingly, the growth of CA5 was not inhibited in the presence of as high as 64 mM selenite. A novel α-proteobacterium, *Rhizobium selenireducens* sp., capable of denitrification and reduction of selenite to elemental selenium under aerobic and denitrifying conditions was isolated from a laboratory bioreactor removing selenate from simulated groundwater. Inhibition of *in vivo* selenite reduction by tungsten suggested involvement of a molybdenum-containing protein in selenite reduction (Hunter et al., 2007a; Hunter et al., 2007b).

B. mycoides and *S. maltophilia* obtained from rhizosphere soil of the selenium hyperaccumulator plant *Astragalus bisulcatus* showed reduction of 67% and 50% of 0.5 mM selenite, respectively, in 48 h, and had even a higher reduction capacity as a consortium (Vallini et al., 2005). Microbial synthesis of CdSe nanoparticles (quantum dots) by the selenite-reducing and cadmium resistant bacterium *Pseudomonas* sp. Strain RB, isolated from soil (used for illegal dumping of electronic appliances), was achieved in the presence of selenite, cadmium ions and cycloheximide as fungal suppressor in aerobic conditions (Ayano et al., 2014). Selenium oxyanions are predominant in aerobic environments (Etezad et al., 2009). In the soil-water environment, aerobic reduction of toxic selenate and selenite by microorganisms to non-toxic elemental selenium can be applied as a bioremediation tool for seleniferous soils. Bioaugmentation of these bacteria in agricultural setting will provide an insight on the nature of plant-microbe interactions and its application in selenium bioremediation (Staicu et al., 2015). Biogenic nanoparticles produced after selenium bioreduction can be used as a potential source of this scarce element for industrial and agricultural applications (Nancharaiah and Lens 2015b).

The ability of the fungus *Phanerochaete chrysosporium* to reduce selenium oxyanions to elemental selenium and the effect of selenium on growth, substrate consumption rate and pellet morphology was studied (Espinosa-Ortiz et al., 2014). 80% fungal growth was inhibited at 10 mg L^{-1} of selenite with 40% selenite removal efficiency under acidic to neutral pH conditions. However, selenate lead to 30% decrease in fungal growth, while less than 10% of 10 mg L^{-1} selenate was removed. Reduction of selenite lead to the intracellular formation of elemental selenium nanoparticles in the range of 30-400 nm (Espinosa-Ortiz et al., 2014).

2.6.3. Anaerobic reduction

Anaerobic reduction of selenite has been observed in a diverse group of microorganisms isolated from pristine and contaminated soils. Under anaerobic conditions, certain microorganisms utilize selenate and selenite as terminal electron acceptors, via dissimilatory reduction to elemental selenium. *Rhodopseudomonas palustris* strain N isolated from a sewage plant exhibited selenite tolerance up to 8.0 mmol L^{-1} concentration and anaerobically reduced selenite to red elemental selenium (Li et al., 2014). The gram positive, rod shaped, facultative anaerobe, endospore forming, nitrate-reducing, motile *Paenibacillus selenitireducens* sp., isolated from selenium mineral soil, reduced selenite to elemental selenium under anaerobic conditions (Yao et al., 2014).

Pearce et al. (2009) studied mechanisms of reduction of selenite to elemental selenium and then selenide in anaerobic conditions by *Geobacter sulfurreducens*, *Shewanella oneidensis* and *Veillonella atypica*. Study of fundamental biochemical mechanisms of selenium reduction in microorganisms is essential in order to develop sustainable selenium bioremediation technologies. The nature of organisms and the prevailing reducing conditions strongly influence the selenite reduction rate. *G. sulfurreducens* completely reduced selenite to selenide in the presence of anthraquinone-2,6-disulfonate (AQDS) as redox mediator, while *S. oneidensis* reduced selenite only upto Se(0). The research attempted to identify the protein-nanoparticle complex formed during biogenesis of the Se(0) nanospheres by *G. sulfurreducens* (Pearce et al., 2009).

2.7. Bioremediation of seleniferous soils

2.7.1. Phytoremediation by use of selenium hyperaccumulators

Phytoremediation has been widely studied as a remediation tool for seleniferous soils because of its cost effectiveness and low-maintenance. Several studies have been carried out to comprehend phytoremediation of seleniferous soils under different conditions. In a lab-based study, Esringü and Turan (2012) reported that addition of synthetic chelating agents such as diethylenetriamine pentaacetate and ethylenediamine disuccinate to selenium contaminated soil assisted in releasing selenium from soil and improved selenium availability for root absorption, thus enhancing selenium removal by 12- to 20-fold by Brussels sprouts.

Suppression of selenium uptake by plants in seleniferous environment is also a promising approach to avoid selenium toxicity. Mackowiak and Amacher (2004) reported amendment of sulfur (either elemental S or gypsum) to selenium enriched soil caused an over 60% suppression of the selenium uptake by Alfalfa and Western Wheatgrass. Sulfate (SO_4^{2-}) being a structural analogue of selenate (SeO_4^{2-}) acts as a competitive anion for uptake by plants and is thus an economically viable option for treating Se-impacted and re-vegetated lands.

Allium cepa, commonly known as onion, accumulated selenium effectively when grown in Se-spiked soil. Onion being an edible component, Yadav et al. (2007) suggested its utilization as a component for selenium mobilization from seleniferous soils to selenium-deficient soils. In a similar study, Yasin et al. (2014) employed the selenium accumulator plant, *Brassica juncea,* to remove Se from seleniferous soils, with and without exogenous bacterial consortium composed of *Bacillus* sp., *Cellulosimicrobium* sp. and *Exiguobacterium* sp. derived from a tannery waste contaminated soil. Although the plant itself demonstrated significant accumulation of selenium in leaves, pod husk and seeds, application of the bacterial consortium further enhanced the efficiency of the process (40-45%) by promoting plant growth.

Lindblom et al. (2014) investigated selenium removal from seleniferous soil by the selenium hyperaccumulator plant *Stanleya pinnata,* and the non-accumulator plant *Stanleya elata* in a greenhouse. Selenium accumulation and speciation in these plants inoculated with the hyperaccumulator rhizospheric fungi- *Alternaria seleniiphila* and *Aspergillus leprosis* was evaluated. Overall, selenium uptake was improved 2 to 3-fold in *S. pinnata* inoculated with *A. seleniiphila* along with a 30% higher selenium uptake in roots compared to uninoculated *S. pinnata.* Those inoculated with *A. leprosis* showed 1.5 times lower selenium accumulation than uninoculated *S. pinnata.* Fungal inoculation showed no significant effect on selenium accumulation in *S. elata,* but negatively affected the plant growth. The finding suggested that inoculation of rhizosphere fungi can affect plant growth and selenium accumulation, depending on the host species. In a greenhouse study, Bañuelos and Lin (2005) evaluated the phytoremediation potential of salt and boron tolerant plant species to clean up Se-laden drainage sediment in San Luis Drain (California, USA). Over the duration of 60 days of experiment, there was a considerable decrease in plant biomass and accumulation of selenium in different plant species in varying concentrations. Although the total selenium content of the soil was lowered by 20%, the water-extractable selenium concentrations increased 3 times to that of the pre-plant concentrations. This was attributed to the influence of plants and irrigation water on biological and physical processes making soluble selenium susceptible to leaching into the groundwater.

2.7.2. Phytoremediation by genetic engineering of plants

Fast growing arabidopsis and Indian mustard plants were genetically engineered to over express the selenocysteine methyltransferase gene from a selenium hyperaccumulator *Astragalus bisulcatus* for phytoremediation of seleniferous soils (Leduc et al., 2004). Selenocysteine methyltransferase specifically catalyses the methylation of SeCys to the non-toxic, non-protein amino acid methylselenocysteine (MetSeCys), thereby reduces the intracellular concentration of organo-Se compounds such as SeCys and SeMet. The transgenic plant showed higher plant biomass productivity and 2 to 4 times more selenium accumulated in the form of methylselenocysteine and increased selenium tolerance. In a similar study, overexpression of the selenocysteine methyltransferase gene from the selenium hyperaccumulator *Astragalus bisulcatus* in a selenium non-accumulator *Arabidopsis thaliana* resulted in significant selenium tolerance and accumulation (Ellis et al., 2004). LeDuc et al. (2006) emphasized the need for development of fast-growing plants, which can tolerate, accumulate and volatilize selenium by use of genetic engineering. ATP sulfurylase is known to catalyse reduction of selenate to organo-Se compounds and is considered a rate-limiting step for the plants over expressing selenocysteine methyltransferase due to stalled selenate reduction. Over expression of ATP sulfurylase and selenocysteine methyltransferase genes in transgenic Indian mustard (*Brassica juncea*) displayed a 4 to 9-fold improvement in selenium tolerance and accumulation compared to the wild-type.

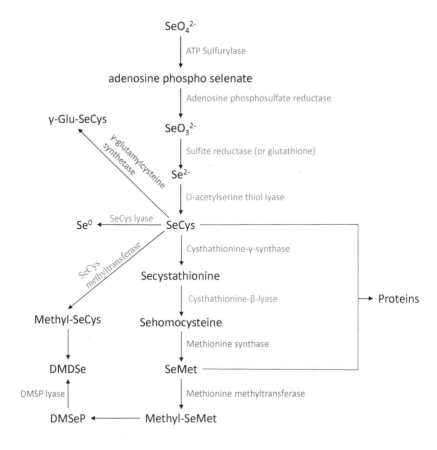

Figure 2.12. Schematic overview of Se metabolism in plants. Reactions catalysed by cytosolic and plastid enzymes are represented in red and blue colour, respectively (Adapted from Pilon-Smits & Quinn, 2010)

In order to avoid selenium toxicity due to nonspecific incorporation of SeCys in proteins instead of cysteine, plants of the genus *Arabidopsis* were genetically engineered to express mouse selenocysteine lyase in cytosol or chloroplast of the plants. Selenocysteine lyase catalyses decomposition of SeCys into non- toxic elemental Se(0) and alanine. Increased tolerance to a high selenium concentration in the environment was observed in plants which expressed selenocysteine lyase (SL) in the cytosol than those that expressed SL in the chloroplast. Selenium accumulation in both transgenic plants was up to 1.5 fold more than that by the wild type (Pilon et al., 2003). Selenocysteine lyase when expressed in *Brassica juncea* resulted in successful, redirection of the selenium flow, away from non-specific incorporation into proteins and upto 50% increase in shoot selenium level at the seedling stage (Garifullina et al., 2003). In a similar study, Van Hoewyk et al. (2005) presented that overexpression of

CpNifS (a chloroplastic NifS-like protein that catalyses the conversion of SeCys to alanine and Se[0]), improved selenate tolerance and accumulation by 1.9 and 2.2-fold, respectively. **Figure 2.12** portrays schematically uptake, metabolism and fate of selenium at enzymatic level in a plant cell.

2.7.3. Microbial remediation by bioaugmentation

The direct and residual effect of applying different organic matter (organic manure, press mud and poultry manure) on selenium accumulation by crops and volatilization from seleniferous soil was studied in greenhouse and field-scale experiments (Dhillon et al. 2010). Utilization of press mud and poultry manure effectively enhanced volatilization and inhibited transfer of selenium from a seleniferous soil to the plants. Fellowes et al. (2013) investigated the effect of microbial activity on selenium cycling in the environment in a Se-enriched soils in Ireland. In microcosm experiments under anoxic conditions, supplementation of soil with selenate resulted in rapid reduction to Se(0). However, addition of sodium acetate as electron donor had no effect on the selenate removal. Rapid denitrification was also observed upon addition of nitrate to the microcosm, hindering selenate reduction and reoxidizing reduced forms of selenium. Thus, the study discourages inclusion of selenate into nitrogenous fertilizers as this could lead to mobilisation and Se-release in surface run-off.

The effect of organic amendments such as poultry manure, sugarcane press mud and farmyard manure on selenium uptake and biochemical grain composition of wheat and rapeseed grown on seleniferous soils in north-western India was investigated by Sharma et al. (2011). This field-based study concluded that application of poultry manure and sugarcane press mud to Se-contaminated soils resulted in a 75-95% decrease in selenium accumulation in both plants (wheat and rapeseed), while farm yard manure amendment decreased selenium accumulation in wheat by 23%, thus avoiding selenium accumulation, alleviating the deleterious effects of selenium and restoring the nutritional quality of the grains.

2.7.4. Microbial remediation by volatilization

Volatile DMSe can be naturally formed from inorganic selenium species in soils. It can regulate the geo-chemical cycling of selenium, thereby influencing selenium bioremediation efficiency. Environmental and soil factors (e.g. organic amendments and frequent tillage) can be optimized to increase diffusive transport through soil and enhance volatilization (Dhillon et al., 2010; Gadd, 2000).

A field-scale experiment was conducted at the Kesterson Reservoir (California, USA) to assess microbial volatilization of selenium as a bioremediation approach to dissipate selenium from contaminated waters (Flury et al. 1997). To enhance microbial methylation of selenium to volatile selenium compounds, the plots filled with contaminated water were amended with different carbon (cattle manure, gluten, orange peel, straw of *Typha latifolia*) and protein (casein) sources and were periodically tilled and irrigated. Soil amendments with different organic and protein sources were hypothesized to improve microbial activity. The highest amount of selenium depletion occurred with amendment of the protein casein, however, no statistically significant difference was observed in selenium removal from the different treatments compared to the case where there was no carbon or protein amendment. In a field study, bioamendment of seleniferous agricultural drainage sediments in San Luis Drain (California, USA) with methionine and casein in vegetated and non-vegetated plots were studied. Rates of selenium volatilization were higher for methionine and casein in vegetated and non-vegetated land, respectively, compared to methionine in non-vegetated land and casein in vegetated land (Bañuelos & Lin, 2007).

In order to provide an insight on radioactive selenium ^{79}Se waste disposal, Ashworth and Shaw (2006) investigated migration, plant uptake and volatilisation of radio-selenium ^{75}Se (surrogate of ^{79}Se) from the contaminated water table in a column experiment using sandy loam soil. Quantification of radio-selenium from the column to soil surface suggested that migration of ^{75}Se through the soil largely depends on the redox potential of soil, with a lower redox potential limiting soil migration. Selenium uptake by plants is observed when roots are in contact with the contaminated soil region, suggesting volatilization of ^{75}Se directly from the soil to the atmosphere, or first uptake by plants followed by its emission to the atmosphere.

2.7.5. Microbial remediation by *in situ* bioreduction of selenium

Low cost microbial reduction to remove selenium with amendment of water bodies with nutrients, including a carbon source, has been effectively applied to reduce the selenium concentration in pit lakes at Sweetwater uranium mine (Wyoming, USA) and the Gilt Edge Superfund Site (South Dakota, USA) (Golder Associates Inc.). Moore et al. (2011) introduced an innovative approach which utilizes a novel chemisorption technology for treatment of selenate using an *in-situ* solidification method by introducing an amorphous sorbent with active sites for chemisorption of selenate and treatment of the contaminated water using this silicate-based inorganic polymer to remove > 99% of the total selenium from groundwater.

Removal of selenium from groundwater was studied by Williams et al. (2013) using injection of acetate into an uranium contaminated aquifer near Rifle, (Colorado, USA). The microbial biofilm formed on the tubing that circulated acetate amended groundwater carried out bioreduction of selenium oxyanions [Se(IV) and Se(VI)] to elemental selenium. 16S rRNA based microbial community analysis revealed coupling of acetate oxidation to reduction of oxygen, nitrate and selenate. *Decholoromonas* sp. and *Thauera* sp. were found to be dominant members of selenium reducing biofilm. *Hydrogenophaga* sp. also constituted about 8% of the biofilm. It was suggested that this microaerophilic bacteria played a role in scavenging oxygen in the biofilm providing complete anaerobic conditions, which enabled selenate and nitrate reduction by *Dechloromonas* sp. and selenate reduction by *Thauera* sp. in the biofilm.

Remediation of selenium-rich soil using *in situ* methods has several disadvantages such as lower rate of remediation, possibilities for groundwater contamination and re-oxidation of insoluble Se(0) to soluble oxyanions. The groundwater is again irrigated on the agricultural field, thus re-circulating the selenium in the same soil-water environment.

2.7.6. *Ex situ* bioremediation by soil washing

Ex situ approaches for seleniferous soil bioremediation have been unexplored in comparison to *in situ* soil bioremediation probably due to their cost factor. One of the common *ex situ* approach applied for remediation of metal-contaminated soils is soil washing for leaching metal contaminants and treatment of leachate solutions (Jiang et al., 2011; Kim et al., 2012). The removal of contaminants from soil using the soil washing is practically an adaptation of mineral processing technologies. Application of soil washing technology has been widely used in Europe for removal of metals like arsenic, cadmium, chromium, copper, mercury, nickel, lead and zinc from contaminated soils (Dermont et al., 2008; Kim et al., 2012; Mohanty and Mahindrakar 2011). Soil washing is considered a pre-concentration stage prior to further treatment of the leachate.

Soil washing may be a useful strategy for remediation of alkaline soils where an appreciable quantity of selenium is present in water-soluble form. With repeated soil extraction using simple reagents like potassium chloride, it was possible to remove up to 90% of the selenium from the seleniferous soils (Dhillon and Dhillon, 2003). Various soil washing steps employed in sequential extraction procedures for characterization of selenium present in different soils can act as a guide for selecting an appropriate soil washing technique for remediation. Soil washing can be combined with anaerobic bioreactors for treatment of leachate and possible recovery of selenium in the form of elemental selenium (Tan et al., 2016). Moreover, this approach can also be explored for the treatment of selenium contaminated soils with a possibility to recover this valuable pollutant for use in technological applications (Mal et al., 2016).

2.8. Conclusions

The selenium content in soil varies greatly throughout the world and its speciation in soil plays a major role in the biogeochemical selenium cycling between the soil-water-atmosphere. Higher amounts of bioavailable forms of selenium in soils greatly influence the amount of selenium in the food chain and selenium levels in higher trophic levels. Elevated selenium content in soil can cause contamination of water bodies and groundwaters due to the leaching caused by rainfall, irrigation and other anthropogenic activities. *In situ* soil bioremediation approaches like phytoremediation using selenium hyper-accumulating plants have demonstrated a promising outlook for removal of selenium from soils using plants. The application of these selenium rich plants as food selenium supplements in selenium deficient populations or fertilizer for selenium deficient agricultural soils has been explored. However, phytoremediation is one of the slowest remediation techniques since it depends on plant growth. Moreover, the mobility and solubility of selenium oxyanions increases its potential to leach out of the soil into the groundwater, thus contaminating it. *In situ* bioremediation approaches such as selenium volatilization and dissimilatory selenate reduction using soil amendment with organic carbon and/or protein have been shown not to improve biotreatment of seleniferous soils. *Ex situ* remediation of seleniferous soils by soil washing followed by biological treatment of the leachate solutions may be a promising alternative for the remediation of contaminated soils coupled to selenium recovery.

References

Ashworth, D. J., Shaw, G., 2006. Soil migration, plant uptake and volatilisation of radio-selenium from a contaminated water table. Sci. Total Environ. 370(2-3), 506–14.

Ayano, H., Miyake, M., Terasawa, K., Kuroda, M., Soda, S., Sakaguchi, T., Ike, M., 2014. Isolation of a selenite-reducing and cadmium-resistant bacterium *Pseudomonas* sp. strain RB for microbial synthesis of CdSe nanoparticles. J. Biosci. Bioeng. 117(5), 576–81.

Bajaj, M., Eiche, E., Neumann, T., Winter, J., Gallert, C., 2011. Hazardous concentrations of selenium in soil and groundwater in North-West India. J. Hazard. Mater. 189(3), 640–6.

Bajaj, M., Winter, J., 2014. Se (IV) triggers faster Te (IV) reduction by soil isolates of heterotrophic aerobic bacteria: formation of extracellular SeTe nanospheres. Microb. Cell Fact. 13(1), 1–10.

Bañuelos, G. S., Arroyo, I., Pickering, I. J., Yang, S. I., Freeman, J. L., 2015. Selenium biofortification of broccoli and carrots grown in soil amended with Se-enriched hyperaccumulator *Stanleya pinnata*. Food Chem. 166, 603–8.

Bañuelos, G. S., Bitterli, C., Schulin, R., 2013. Fate and movement of selenium from drainage sediments disposed onto soil with and without vegetation. Environ. Pollut. 180, 7–12.

Bañuelos, G. S., Lin, Z. Q., 2007. Acceleration of selenium volatilization in seleniferous agricultural drainage sediments amended with methionine and casein. Environ. Pollut. 150(3), 306–312.

Bañuelos, G. S., Lin, Z.-Q., 2005. Phytoremediation management of selenium-laden drainage sediments in the San Luis Drain: a greenhouse feasibility study. Ecotoxicol. Environ. Saf. 62(3), 309–16.

Bañuelos, G. S., Mayland, H. F., 2000. Absorption and distribution of selenium in animals consuming canola grown for selenium phytoremediation. Ecotoxicol. Environ. Saf. 46(3), 322–8.

Bodnar, M., Szczyglowska, M., Konieczka, P., Namiesnik, J., 2016. Methods of selenium supplementation: bioavailability and determination of selenium compounds. Crit. Rev. Food Sci. Nutr. 56(1), 36–55.

Brown, K. M., Arthur, J. R., 2001. Selenium, selenoproteins and human health: a review. Public Health Nutr. 4(2B), 593–599.

Butterman, W. C., Brown, R. D., (2004). Selenium. Mineral commodity profile report 03-018, U. S. Geological Survey.

Chen, J., 2012. An original discovery: selenium deficiency and Keshan disease (an endemic heart disease). Asia Pac. J. Clin. Nutr. 21(3), 320–326.

Chen, J. J., Boylan, L. M., Wu, C. K., Spallholz, J. E., 2007. Oxidation of glutathione and superoxide generation by inorganic and organic selenium compounds. BioFactors. 31(1), 55–66.

Clark, S. K., Johnson, T. M., 2010. Selenium stable isotope investigation into selenium biogeochemical cycling in a lacustrine environment: Sweitzer Lake, Colorado. J. Environ. Qual. 39(6), 2200 – 2210.

Curtin, D., Hanson, R., Lindley, T. N., Butler, R. C., 2006. Selenium concentration in wheat (*Triticum aestivum*) grain as influenced by method, rate, and timing of sodium selenate application. New Zeal. J. Crop Hort. 34(4), 329–339.

Cuvardic, M. S., 2003. Selenium in Soil. Proceedings for Natural Sciences, Matica Srpska Novi Sad, 104, 23–37.

De La Riva, D. G., Vindiola, B. G., Castañeda, T. N., Parker, D. R., Trumble, J. T., 2014. Impact of selenium on mortality, bioaccumulation and feeding deterrence in the invasive Argentine ant, Linepithema humile (*Hymenoptera: Formicidae*). Sci. Total Environ. 481, 446–52.

Dermont, G., Bergeron, M., Mercier, G., Richer-Laflèche, M., 2008. Soil washing for metal removal: A review of physical/chemical technologies and field applications. J. Hazard. Mater. 152(1), 1–31.

Dhillon, K. S., & Dhillon, S. K., 2003. Distribution and management of seleniferous soils in: Advances in Agronomy, 79, 119–184.

Dhillon, K.S., Dhillon, S.K., 2014. Development and mapping of seleniferous soils in northwestern India. Chemosphere 99, 56–63.

Dhillon, K.S., Dhillon, S.K., Dogra, R., 2010. Selenium accumulation by forage and grain crops and volatilization from seleniferous soils amended with different organic materials. Chemosphere 78, 548–56.

Diamond, A., 2015. The subcellular location of selenoproteins and the impact on their function. Nutrients 7, 3938–3948.

Dowdle, P.R., Oremland, R.S., 1998. Microbial oxidation of elemental selenium in soil slurries and bacterial cultures. Environ. Sci. Technol. 32, 3749–3755.

Downs, R.T. and Hall-Wallace, M. (2003). The American mineralogist crystal structure database. American Mineralogist 88, 247-250.

Dungan, R. S., Yates, S. R., & Frankenberger, W. T. (2003). Transformations of selenate and selenite by *Stenotrophomonas maltophilia* isolated from a seleniferous agricultural pond sediment. Environmental Microbiology, 5(4), 287–295.

Duntas, L.H., Benvenga, S., 2014. Selenium: an element for life. Endocrine 48, 756–775.

El-Aassar S.A., Berekaa M.M., EL-Shaer M., Y.G.A. and S.J.F., 2002. Microbial transformation of arsenate and selenate in the Nile delta, in: 4th European Bioremediation Conference. 1–12.

Ellis, D.R., Sors, T.G., Brunk, D.G., Albrecht, C., Orser, C., Lahner, B., Wood, K. V, Harris, H.H., Pickering, I.J., Salt, D.E., 2004. Production of Se-methylselenocysteine in transgenic plants expressing selenocysteine methyltransferase. BMC Plant Biol. 4, 1.

Ellwood, M.J., Schneider, L., Potts, J., Batley, G.E., Floyd, J., Maher, W.A., 2016. Volatile selenium fluxes from selenium-contaminated sediments in an Australian coastal lake. Environ. Chem. 13, 68.

El-Ramady, H., Abdalla, N., Alshaal, T., Domokos-Szabolcsy, É., Elhawat, N., Prokisch, J., Sztrik, A., Fári, M., El-Marsafawy, S., Shams, M.S., 2014. Selenium in soils under climate change, implication for human health. Environ. Chem. Lett.13(1), 1-19.

Espinosa-Ortiz, E.J., Gonzalez-Gil, G., Saikaly, P.E., van Hullebusch, E.D., Lens, P.N.L., 2014. Effects of selenium oxyanions on the white-rot fungus *Phanerochaete chrysosporium*. Appl. Microbiol. Biotechnol. 99, 2405–18.

Esringü, A., Turan, M., 2012. The roles of diethylenetriamine pentaacetate (DTPA) and ethylenediamine disuccinate (EDDS) in remediation of selenium from contaminated soil by brussels sprouts (*Brassica oleracea* var. *Gemmifera*). Water. Air. Soil Pollut. 223, 351–362.

Expert group on vitamins and minerals. 2003. Safe upper levels of vitamins and minerals. Food Safety Agency.

Etezad, S.M., Khajeh, K., Soudi, M., Ghazvini, P.T.M., Dabirmanesh, B., 2009. Evidence on the presence of two distinct enzymes responsible for the reduction of selenate and tellurite in *Bacillus* sp. STG-83. Enzyme Microb. Technol. 45, 1–6.

Fairweather-tait, S.J., Collings, R., Hurst, R., 2010. Selenium bioavailability: current knowledge and future research. Am. J. Clin. Nutr. 91, 1484S–91S.

Fellowes, J.W., Pattrick, R. A. D., Boothman, C., Al Lawati, W.M.M., van Dongen, B.E., Charnock, J.M., Lloyd, J.R., Pearce, C.I., 2013. Microbial selenium transformations in seleniferous soils. Eur. J. Soil Sci. 64, 629–638.

Flury, M., Jr, W.T.F., Jury, W.A., 1997. Long-term depletion of selenium from Kesterson dewatered sediments 198, 259–270.

Fordyce, F., 2005. Selenium deficiency and toxicity in the environment. Essentials Med. Geol. 375–416.

Fordyce, F., 2007. Selenium geochemistry and health. AMBIO A J. Hum. Environ. 36, 94–97.

Foster, C.B., 2005. Selenoproteins and the metabolic features of the archaeal ancestor of eukaryotes. Mol. Biol. Evol. 22, 383–386.

Frankenberger, Jr., W.T., Amrhein, C., Fan, T.W.M., Flaschi, D., Glater, J., Kartinen, Jr., E., Kovac, K., Lee, E., Ohlendorf, H.M., Owens, L., Terry, N., Toto, A., 2004. Advanced treatment technologies in the remediation of seleniferous drainage waters and sediments. Irrig. Drain. Syst. 18, 19–42.

Gadd, G.M., 2000. Bioremedial potential of microbial mechanisms of metal mobilization and immobilization. Curr. Opin. Biotechnol. 11, 271–279.

Garbisu, C., Carlson, D., Adamkiewicz, M., Yee, B.C., Wong, J.H., Resto, E., Leighton, T., Buchanan, B.B., 1999. Morphological and biochemical responses of *Bacillus subtilis* to selenite stress. Biofactors 10, 311–319.

Garifullina, G.F., Owen, J.D., Lindblom, S.-D., Tufan, H., Pilon, M., Pilon-Smits, E. A. H., 2003. Expression of a mouse selenocysteine lyase in *Brassica juncea* chloroplasts affects selenium tolerance and accumulation. Physiol. Plant. 118, 538–544.

Gladyshev, V.,2006. Selenoproteins and selenoproteomes. In Selenium: Its molecular biology and role in human health, Second Edition. 99–110.

Goh, K.-H., Lim, T.-T., 2004. Geochemistry of inorganic arsenic and selenium in a tropical soil: effect of reaction time, pH, and competitive anions on arsenic and selenium adsorption. Chemosphere 55, 849–59.

Golder Associates Inc., 2009. Literature review of trreatment technologies to remove selenium from mining influenced water.

Hagarova, I., Zemberyova, M., & Bajcan, D. (2003). Sequential and single step extraction procedures for fractionation of selenium in soil samples. Chemical Papers, 59(2), 93–98.

Herbel, M. J., Johnson, T. M., Tanji, K. K., Gao, S., & Bullen, T. D. (2002). Selenium stable isotpe rations in California agricultural drainage water management systems. Journal of Environmental quality, 31, 1146–1156.

Hladun, K. R., Parker, D. R., Tran, K. D., & Trumble, J. T. (2013). Effects of selenium accumulation on phytotoxicity, herbivory, and pollination ecology in radish (*Raphanus sativus* L.). Environmental Pollution, 172, 70–5.

Hu, J., Zhao, Q., Cheng, X., Selomulya, C., Bai, C., Zhu, X.,Li, X. & Xiong, H. (2014). Antioxidant activities of Se-SPI produced from soybean as accumulation and biotransformation reactor of natural selenium. Food Chemistry, 146, 531–7.

Huawei, Z. (2009). Selenium as an essential micronutrient: Roles in cell cycle and apoptosis. Molecules, 14(3), 1263–1278.

Hunter, W. J., & Kuykendall, L. D. (2007). Reduction of selenite to elemental red selenium by Rhizobium sp. strain B1. Current Microbiology, 55(4), 344–9.

Hunter, W. J., Kuykendall, L. D., & Manter, D. K. (2007). *Rhizobium selenireducens* sp. nov.: a selenite-reducing *α-Proteobacteria* isolated from a bioreactor. Current Microbiology, 55(5), 455–60.

Hunter, W. J., & Manter, D. K. (2008). Bio-reduction of selenite to elemental red selenium by *Tetrathiobacter kashmirensis*. Current Microbiology, 57(1), 83–88.

Hunter, W. J., & Manter, D. K. (2009). Reduction of selenite to elemental red selenium by *Pseudomonas* sp. Strain CA5. Current Microbiology, 58(5), 493–8.

Ikram, M., & Faisal, M. (2010). Comparative assessment of selenite (SeIV) detoxification to elemental selenium (Se0) by *Bacillus* sp. Biotechnology Letters, 32(9), 1255–1259.

Jiang, W., Tao, T., & Liao, Z. (2011). Removal of heavy metal from contaminated soil with chelating Agents. Open Journal of Soil Science, 1(2), 71–77.

Kabata-pendias, A., Pendias, H., 2001. Trace elements in soils and plants trace elements in soils and plants, 3rd ed. CRC press LLC, Boca Raton, Florida.

Kalender, S., Uzun, F. G., Demir, F., Uzunhisarcikli, M., & Aslanturk, A. (2013). Mercuric chloride-induced testicular toxicity in rats and the protective role of sodium selenite and vitamin E. Food and Chemical Toxicology, 55, 456–462.

Kaur, N., Sharma, S., & Nayyar, H. (2014). Selenium in agriculture: A nutrient or toxin for crops? Archives of Agronomy and Soil Science, 60 (12)(October), 1593–1624.

Kausch, M., Ng, P., Ha, J., & Pallud, C. (2012). Soil-aggregate-scale heterogeneity in microbial selenium reduction. Vadose Zone Journal, 11(2).

Keskinen, R., Ekholm, P., Yli-Halla, M., & Hartikainen, H. (2009). Efficiency of different methods in extracting selenium from agricultural soils of Finland. Geoderma, 153(1-2), 87–93.

Kessi, J., & Hanselmann, K. W. (2004). Similarities between the abiotic reduction of selenite with glutathione and the dissimilatory reaction mediated by *Rhodospirillum rubrum* and *Escherichia coli*. Journal of Biological Chemistry, 279(49), 50662–50669.

Kim, K., Cheong, J., Kang, W., Chae, H., & Chang, C. (2012). Field study on application of soil washing system to arsenic- contaminated site adjacent to J. refinery in Korea. In International Conference on Environmental Science and Technology, 30, 1–5.

Kryukov, G. V, & Gladyshev, V. N. (2004). The prokaryotic selenoproteome. EMBO Reports, 5(5), 538–543.

Kulp, T. R., & Pratt, L. M. (2004). Speciation and weathering of selenium in upper cretaceous chalk and shale from South Dakota and Wyoming, USA. Geochimica et Cosmochimica Acta, 68(18), 3687–3701.

Kurokawa, S., & Berry, M. J. (2013). Selenium. Role of the essential metalloid in health. Metal Ions in Life Science, 13(34), 499–534.

Laird, K. R., Das, B., & Cumming, B. F. (2014). Enrichment of uranium, arsenic, molybdenum, and selenium in sediment cores from boreal lakes adjacent to northern Saskatchewan uranium mines. Lake and Reservoir Management, 30(4), 344–357.

Lampis, S., Zonaro, E., Bertolini, C., Bernardi, P., Butler, C. S., & Vallini, G. (2014). Delayed formation of zero-valent selenium nanoparticles by *Bacillus mycoides* SeITE01 as a consequence of selenite reduction under aerobic conditions. Microbial Cell Factories, 13(1), 35.

Lavu, R. V. S., Du Laing, G., Van De Wiele, T., Pratti, V. L., Willekens, K., Vandecasteele, B., & Tack, F. (2012). Fertilizing soil with selenium fertilizers: Impact on concentration, speciation, and bioaccessibility of selenium in leek (*Allium ampeloprasum*). Journal of Agricultural and Food Chemistry, 60(44), 10930–10935.

LeDuc, D. L., AbdelSamie, M., Móntes-Bayon, M., Wu, C. P., Reisinger, S. J., & Terry, N. (2006). Overexpressing both ATP sulfurylase and selenocysteine methyltransferase enhances selenium phytoremediation traits in Indian mustard. Environmental Pollution, 144(1), 70–76.

Leduc, D. L., Tarun, A. S., Montes-bayon, M., Meija, J., Malit, M. F., Wu, C. P., Abdelsamie, M., Chiang, C., Tagmount, A., Neuhierl, B., Caruso, J. & Terry, N. (2004). Overexpression of selenocysteine methyltransferase in Arabidopsis and Indian mustard increases selenium tolerance and accumulation. Plant Physiology, 135(May), 377–383.

Lemly, a. D. (2014). Teratogenic effects and monetary cost of selenium poisoning of fish in Lake Sutton, North Carolina. Ecotoxicology and Environmental Safety, 104(1), 160–167.

Lenz, M. (2008). Biological selenium removal from wastewaters. [Doctoral thesis]. Wageningen University, Wageningen, the Netherlands

Lenz, M., & Lens, P. N. L. (2009). The essential toxin: the changing perception of selenium in environmental sciences. The Science of the Total Environment, 407(12), 3620–33.

Li, B., Liu, N., Li, Y., Jing, W., Fan, J., Li, D., Zhang, L., Zhang, X., Zhang, Z. & Wang, L. (2014). Reduction of selenite to red elemental selenium by *Rhodopseudomonas palustris* strain N. PloS One, 9(4), e95955.

Li, H. F., McGrath, S. P., & Zhao, F. J. (2008). Selenium uptake, translocation and speciation in wheat supplied with selenate or selenite. New Phytologist, 178(1), 92–102.

Lindblom, S. D., Fakra, S. C., Landon, J., Schulz, P., Tracy, B., & Pilon-Smits, E. A. H. (2014). Inoculation of selenium hyperaccumulator *Stanleya pinnata* and related non-accumulator *Stanleya elata* with hyperaccumulator rhizosphere fungi - investigation of effects on Se accumulation and speciation. Physiologia Plantarum, 150(1), 107–18.

Lyons, G. (2010). Selenium in cereals: Improving the efficiency of agronomic biofortification in the UK. Plant and Soil, 332(1), 1–4.

Mackowiak, C. L., & Amacher, M. C. (2007). Soil sulfur amendments suppress selenium uptake by alfalfa and Western wheatgrass. Journal of Environmental Quality, 37(3), 772–9.

Mal, J., Nancharaiah, Y. V., van Hullebusch, E. D., & Lens, P. N. L. (2016). Metal Chalcogenide quantum dots: biotechnological synthesis and applications. RSC Advances, 6(April), 41477–4149.

Martens, D. A. (1997). Selenium speciation of soil/sediment determined with sequential extractions and hydride generation atomic absorption spectrophotometry selenium speciation of soil/sediment determined with sequential extractions and hydride. Environmental Science & Technology, 31, 133-139

Mayland, H. F., James, L. F., Panter, K. E., & Sonderegger, J. L. (1989). Selenium in seleniferous environments. Selenium in Agriculture and the Environment, (23), 15–50.

Mehdi, Y., Hornick, J.-L., Istasse, L., & Dufrasne, I. (2013). Selenium in the environment, metabolism and involvement in body functions. Molecules (Basel, Switzerland), 18(3), 3292–311.

Meseck, S., & Cutter, G. (2012). Selenium behavior in San Francisco bay sediments. Estuaries and Coasts, 35(2), 646–657.

Miller, L. L., Rasmussen, J. B., Palace, V. P., Sterling, G., & Hontela, A. (2013). Selenium bioaccumulation in stocked fish as an indicator of fishery potential in pit lakes on reclaimed coal mines in Alberta, Canada. Environmental Management, 52(1), 72–84.

Mirza Hasanuzzaman, M. Anwar Hossain, M. F. (2010). Selenium in higher plants: Physiological role, antioxidant metabolism and abiotic stress tolerance. Journal of Plant Sciences, 5 (4), 354–375.

Mishra, R. R., Prajapati, S., Das, J., Dangar, T. K., Das, N., & Thatoi, H. (2011). Reduction of selenite to red elemental selenium by moderately halotolerant *Bacillus megaterium* strains isolated from Bhitarkanika mangrove soil and characterization of reduced product. Chemosphere, 84(9), 1231–7.

Misra, S., Boylan, M., Selvam, A., Spallholz, J., & Björnstedt, M. (2015). Redox-active selenium compounds—From toxicity and cell death to cancer treatment. Nutrients, 7(5), 3536–3556.

Mohanty, B., & Mahindrakar, A. B. (2011). Removal of heavy metal by screening followed by soil washing from contaminated soil. International Journal of Technology and Engineering Systems, 2(3), 290–293.

Moore, L., & Mahmoudkhani, A. (2011). Methods for removing selenium from aqueous systems. Proceedings Tailings and Mine Waste.

Moreno, R. G., Burdock, R., Cruz, M., Álvarez, D., & Crawford, J. W. (2013). Managing the selenium content in soils in semiarid environments through the recycling of organic matter. Applied and Enivironmental Soil Science, 2013, 1–10.

Nancharaiah, Y. V., Mohan, S. V., & Lens, P. N. L. (2016). Biological and bioelectrochemical recovery of critical and scarce metals. Trends in Biotechnology, 34(2), 137–155.

Nancharaiah, Y. V, & Lens, P. N. L. (2015a). The ecology and biotechnology of selenium respiring bacteria. Microbiology and Molecular Biology Reviews, 79(1), 61–80.

Nancharaiah, Y.V., Lens, P.N.L., (2015b). Selenium biomineralization for biotechnological applications. Trends in Biotechnolology 33(6), 323-330.

Naumov, A. V. (2010). Selenium and tellurium: State of the markets, the crisis, and its consequences. Metallurgist, 54(3-4), 197–200.

Navarro-Alarcon, M., & Cabrera-Vique, C. (2008). Selenium in food and the human body: A review. Science of the Total Environment, 400(1-3), 115–141.

Oldfield, J. E. (2002). Selenium World atlas. Selenium-tellurium development association.

Papp, L. V., Lu, J., Holmgren, A., & Khanna, K. K. (2007). From selenium to selenoproteins: synthesis, identity, and their role in human health. Antioxidants & Redox Signaling, 9(7), 775–806.

Parkman, H., & Hultberg, H. (2002). Occurence and effects of selenium in the environment - a literature review.

Pearce, C. I., Pattrick, R. a D., Law, N., Charnock, J. M., Coker, V. S., Fellowes, J. W., Oremland, R. S. & Lloyd, J. R. (2009). Investigating different mechanisms for biogenic selenite transformations: *Geobacter sulfurreducens*, *Shewanella oneidensis* and *Veillonella atypica*. Environmental Technology, 30(12), 1313–26.

Peng, Q., Guo, L., Ali, F., Li, J., Qin, S., Feng, P., & Liang, D. (2016). Influence of Pak choi plant cultivation on Se distribution, speciation and bioavailability in soil. Plant and Soil. 403(1), 331-342

Pilon, M., Owen, J. D., Garifullina, G. F., Kurihara, T., Mihara, H., Esaki, N., & Pilon-Smits, E. A. H. (2003). Enhanced selenium tolerance and accumulation in transgenic Arabidopsis expressing a mouse selenocysteine lyase. Plant Physiology, 131(3), 1250–1257.

Pilon-Smits, E. A. H., Quinn, C. F., Tapken, W., Malagoli, M., & Schiavon, M. (2009). Physiological functions of beneficial elements. Current Opinion in Plant Biology, 12(3), 267–74.

Pilon-Smits, E. A. H., & Quinn, C. F. (2010). Selenium metabolism in plants. In Cell Biology of Metals and Nutrients (Vol. 17, pp. 281–298).

Ponce de Leon, C. A., DeNicola, K., Bayon, M. M., & Caruso, J. A. (2003). Sequential extractions of selenium soils from Stewart Lake: total selenium and speciation measurements with ICP-MS detection. Journal of Environmental Monitoring, 5(3), 435–440.

Rautiainen, S., Manson, J. E., Lichtenstein, A. H., & Sesso, H. D. (2016). Dietary supplements and disease prevention - a global overview. Nature Reviews Endocrinology, 12(7), 407-420

Rayman, M. P. (2012). Selenium and human health. The Lancet, 379(9822), 1256–1268.

Rayman, M. P., Infante, H. G., & Sargent, M. (2008). Food-chain selenium and human health: spotlight on speciation. The British Journal of Nutrition, 100(2), 238–253.

Ros, G. H., van Rotterdam, A. M. D., Bussink, D. W., & Bindraban, P. S. (2016). Selenium fertilization strategies for bio-fortification of food: an agro-ecosystem approach. Plant and Soil, 404(1-2), 99-112.

Saidi, I., Chtourou, Y., & Djebali, W. (2014). Selenium alleviates cadmium toxicity by preventing oxidative stress in sunflower (*Helianthus annuus*) seedlings. Journal of Plant Physiology, 171(5), 85–91.

Sarret, G., Avoscan, L., Carrie, M., Collins, R., Geoffroy, N., Carrot, F., Cove`s, J. & Gouget, B. (2005). Chemical forms of selenium in the metal-resistant bacterium *Ralstonia metallidurans* CH34 exposed to selenite and selenate. Applied and Environmental Microbiology, 71(5), 2331–2337.

Schilling, K., Johnson, T. M., Dhillon, K. S., & Mason, P. R. D. (2015). Fate of selenium in soils at a seleniferous site recorded by high precision Se isotope measurements. Environmental Science and Technology, 49(16), 9690–9698.

Selenius, M., Rundlöf, A.-K., Olm, E., Fernandes, A. P., & Björnstedt, M. (2010). Selenium and the selenoprotein thioredoxin reductase in the prevention, treatment and diagnostics of cancer. Antioxidants & Redox Signaling, 12(7), 867–880.

Sharma, S., Bansal, A., Dogra, R., Dhillon, S. K., & Dhillon, K. S. (2011). Effect of organic amendments on uptake of selenium and biochemical grain composition of wheat and rape grown on seleniferous soils in northwestern India. Journal of Plant Nutrition and Soil Science, 174(2), 269–275.

Staicu, L. C., Ackerson, C. J., Cornelis, P., Ye, L., Berendsen, R. L., Hunter, W. J., Noblitt, S. D., Henry, C. S., Cappa, J. J., Montenieri, R. L., Wong, A. O., Musilova, L., Sura-de Jong, M., van Hullebusch, E. D., Lens, P. N. L., Reynolds, R. J. B. & Pilon-Smits, E. A. H. (2015). *Pseudomonas moraviensis* subsp. stanleyae, a bacterial endophyte of hyperaccumulator *Stanleya pinnata*, is capable of efficient selenite reduction to elemental selenium under aerobic conditions. Journal of Applied Microbiology, 119(2), 400–410.

Statwick, J., Majestic, B. J., & Sher, A. A. (2016). Characterization and benefits of selenium uptake by an Astragalus hyperaccumulator and a non-accumulator. Plant and Soil. (In press)

Stock, T., & Rother, M. (2009). Selenoproteins in Archaea and Gram-positive bacteria. Biochimica et Biophysica Acta - General Subjects, 1790(11), 1520–1532.

Stolz, J. F., & Oremland, R. S. (1999). Bacterial respiration of arsenic and selenium. FEMS Microbiology Reviews, 23(5), 615–627.

Sun, G.-X., Meharg, A. A., Li, G., Chen, Z., Yang, L., Chen, S.-C., & Zhu, Y.-G. (2016). Distribution of soil selenium in China is potentially controlled by deposition and volatilization? Scientific Reports, 6(October 2015), 20953.

Tamaoki, M., Freeman, J. L., Pilon-smits, E. A. H., Collins, F., & T, C. M. (2008). Cooperative Ethylene and Jasmonic Acid Signaling Regulates Selenite Resistance in Arabidopsis. Plant Physiology, 146(March), 1219–1230.

Tan, L. C., Nancharaiah, Y. V., van Hullebusch, E. D., & Lens, P. N. L. (2016). Selenium: Environmental significance, pollution, and biological treatment technologies. Biotechnology Advances, 34(5), 886-907.

Tapiero, H., Townsend, D. M., & Tew, K. D. (2003). The antioxidant role of selenium and seleno-compounds. Biomedicine and Pharmacotherapy, 57(3), 134–144.

Tinggi, U. (2003). Essentiality and toxicity of selenium and its status in Australia : a review. Toxicology Letters, 137, 103–110.

Tinggi, U. (2008). Selenium: Its role as antioxidant in human health. Environmental Health and Preventive Medicine, 13(2), 102–108.

Tolu, J., Thiry, Y., Bueno, M., Jolivet, C., Potin-Gautier, M., & Le Hécho, I. (2014). Distribution and speciation of ambient selenium in contrasted soils, from mineral to organic rich. The Science of the Total Environment, 479-480, 93–101.

Tugarova, A. V, Vetchinkina, E. P., Loshchinina, E. a, Burov, A. M., Nikitina, V. E., & Kamnev, A. a. (2014). Reduction of selenite by Azospirillum brasilense with the formation of selenium nanoparticles. Microbial Ecology, 68(3), 495–503.

Turner, A. (2013). Selenium in sediments and biota from estuaries of southwest England. Marine Pollution Bulletin, 73(1), 192–198.

Vallini, G., Gregorio, S. Di, & Lampis, S. (2005). Rhizosphere-induced Selenium Precipitation for Possible Applications in Phytoremediation of Se Polluted Effluents. Z. Naturforsch.

60c, 349-356.

Van Hoewyk, D., Garifullina, G. F., Ackley, A. R., Abdel-Ghany, S. E., Marcus, M. a, Fakra, S., Ishiyama, K., Inoue, E., Pilon, M., Takahashi, H. & Pilon-Smits, E. A. H. (2005). Overexpression of AtCpNifS enhances selenium tolerance and accumulation in Arabidopsis. Plant Physiology, 139(3), 1518–1528.

Wang, J., Li, H., Yang, L., Li, Y., Wei, B., Yu, J., & Feng, F. (2016). Distribution and translocation of selenium from soil to highland barley in the Tibetan plateau Kashin-Beck disease area. Environmental Geochemistry and Health. 1-12

Wang, Q., Zhang, J., Zhao, B., Xin, X., Deng, X., & Zhang, H. (2016). Influence of long-term fertilization on selenium accumulation in soil and uptake by crops. Pedosphere, 26(1), 120–129.

Wang, Y., Dang, F., Zhao, J., & Zhong, H. (2016). Selenium inhibits sulfate-mediated methylmercury production in rice paddy soil. Environmental Pollution, 213, 232–239.

Williams, K. H., Wilkins, M. J., N'Guessan, a. L., Arey, B., Dodova, E., Dohnalkova, A., Holmes, D., Lovley, D. R. & Long, P. E. (2013). Field evidence of selenium bioreduction in a uranium-contaminated aquifer. Environmental Microbiology Reports, 5(3), 444–452.

Winkel, L., Feldmann, J., & Meharg, A. a. (2010). Quantitative and qualitative trapping of volatile methylated selenium species entrained through nitric acid. Environmental Science & Technology, 44(1), 382–7.

Winkel, L. H. E., Johnson, C. A., Lenz, M., Grundl, T., Leupin, O. X., Amini, M., & Charlet, L. (2012). Environmental selenium research: from microscopic processes to global understanding. Environmental Science & Technology, 46(2), 571–9.

Winkel, L., Vriens, B., Jones, G., Schneider, L., Pilon-Smits, E., & Bañuelos, G. (2015). Selenium Cycling Across Soil-Plant-Atmosphere Interfaces: A Critical Review. Nutrients, 7(6), 4199–4239.

World Health Organization. (2011). Selenium in Drinking-water. World Health Organization, Geneva.

Wright, M. T., Parker, D. R., & Amrhein, C. (2003). Critical Evaluation of the Ability of Sequential Extraction Procedures To Quantify Discrete Forms of Selenium in Sediments and Soils. Environmental Science & Technology, 37(20), 4709–4716.

Wu, L. (2004). Review of 15 years of research on ecotoxicology and remediation of land contaminated by agricultural drainage sediment rich in selenium. Ecotoxicology and Environmental Safety, 57(3), 257–69.

Yadav, S., Gupta, S., Prakash, R., Spallholz, J., & Prakash, N. T. (2007). Selenium uptake by Allium cepa grown in Se-spiked soils. American-Eurasian Journal of Agrucultural and Environmental Science, 2(1), 80-84.

Yao, R., Wang, R., Wang, D., Su, J., Zheng, S., & Wang, G. (2014). *Paenibacillus selenitireducens* sp. nov., a selenite-reducing bacterium isolated from a selenium mineral soil. International Journal of Systematic and Evolutionary Microbiology, 64(3), 805–11.

Yasin, M., El Mehdawi, A. F., Jahn, C. E., Anwar, A., Turner, M. F. S., Faisal, M., & Pilon-Smits, E. A. H. (2014). Seleniferous soils as a source for production of selenium-enriched foods and potential of bacteria to enhance plant selenium uptake. Plant and Soil. 386(1), 385-394.

Yu, T., Yang, Z., Lv, Y., Hou, Q., Xia, X., Feng, H., Zhang, M., Jin, L. & Kan, Z. (2014). The origin and geochemical cycle of soil selenium in a Se-rich area of China. Journal of Geochemical Exploration, 139, 97–108.

Yu, X. Z., & Gu, J. D. (2008). Differences in uptake and translocation of selenate and selenite by the weeping willow and hybrid willow. Environmental Science and Pollution Research, 15(6), 499–508.

Zhang, W., Chen, Z., Liu, H., Zhang, L., Gao, P., & Li, D. (2011). Biosynthesis and structural characteristics of selenium nanoparticles by *Pseudomonas alcaliphila*. Colloids and Surfaces. B, Biointerfaces, 88(1), 196–201.

Zhang, Y., & Moore, J. N. (1996). Selenium fractionation and speciation in a wetland system. Environmental Science & Technology, 30(8), 2613–2619.

Zhang, Y., Okeke, B. C., & Frankenberger, W. T. (2008). Bacterial reduction of selenate to elemental selenium utilizing molasses as a carbon source. Bioresource Technology, 99(5), 1267–73.

Zhang, Y., Romero, H., Salinas, G., & Gladyshev, V. N. (2006). Dynamic evolution of selenocysteine utilization in bacteria: a balance between selenoprotein loss and evolution of selenocysteine from redox active cysteine residues. Genome Biology, 7(10), R94.

CHAPTER 3

Optimisation of soil washing for seleniferous soil from Northern India

This chapter has been modified and published as:

Wadgaonkar SL, Ferraro A, Race M, Nancharaiah YV, Dhillon KS, Fabbricino M, Esposito G, Lens PNL. (2018) Optimisation of soil washing to reduce the selenium levels of seleniferous soil from Punjab, Northwestern India. J. Environ. Qual. DOI: 10.2134/jeq2018.05.0187

Abstract

Seleniferous soil collected from the wheat-grown agricultural land in Punjab (India) was characterized for pH, organic matter, moisture content, cation exchange capacity, bulk density, heavy metals and total Se concentration. Further, the Se concentration in various soil fractions was determined through two different sequential extraction procedures. The soil had a total Se content of 4.75 ($\pm$0.02) mg Se kg^{-1}, of which 45% was observed in the oxidisable soil fraction. Both *in situ* and *ex situ* techniques were involved for Se contaminated soil remediation and process efficiency comparison. Soil flushing process was involved as an *in situ* technique and performed to simulate Se migration pattern in case of rainfall or irrigation. A lower Se removal efficiency with increasing layers of the soil column was observed. Also, significant migration of Se from the upper layer to the lower layers was observed during water percolation through the soil column. For *ex situ* treatment, the soil washing technique was optimized by varying different parameters such as treatment time, temperature, pH and liquid to solid (L:S) ratio. The soil washing extraction efficiency of Se was further evaluated in the presence of competing ions and oxidizing agents. Results showed that Se extraction from soil was significantly improved by the presence of oxidizing agents in the washing solution. Around 38% Se was removed from the soil in the presence of 0.5% $KMnO_4$. In contrast, other parameters such as treatment time, temperature, pH, L:S ratio and competing ions showed no significant enhancement of the Se extraction efficiency.

Key words: Seleniferous soil, sequential extraction, soil washing, soil flushing, selenium removal, selenium migration.

3.1. Introduction

Selenium (Se) is a redox-sensitive trace element essential for animals and humans and plays an important role in redox regulation of intracellular signalling, redox homeostasis and thyroid hormone metabolism (El-Ramady et al., 2015). Due to its tendency to bioaccumulate and bio-magnify with higher tropic levels in the food chain (Wu, 2004), the Se content in a soil is a direct source of nutrients to human and animals. Thus, Se in soils directly affects the Se concentration in the food chain, possibly leading to either deficiency or toxicity, both of which have been known to cause several physiological dysfunctions (Lenz and Lens, 2009; Tinggi, 2003). Therefore, determination of the Se content in soils is necessary (Hagarova et al., 2003) as well as the implementation of suitable soil remediation techniques to deal with Se toxicity.

Nowadays, several *in situ* techniques such as phytoremediation (Gupta and Gupta, 2017; Schiavon and Pilon-Smits, 2017), bioaugmentation (Kirk, 2014), bioamendment (Mackowiak and Amacher, 2007; Sharma et al., 2011) and volatilisation (Ashworth and Shaw, 2006) approaches have been applied for bioremediation of seleniferous soils. These techniques aim for bioreduction of toxic Se oxyanions to non-toxic, insoluble elemental Se. Although *in situ* bioremediation techniques provide a clean and green solution, they are slow processes and pose a high risk of groundwater contamination and re-oxidation of Se (Wadgaonkar et al., *In press*). Also, Se being a scarce and critical element (Nancharaiah et al., 2016) is lost during processes such as volatilisation.

Implementation of *ex situ* techniques for clean-up of seleniferous soils have been suggested. Dhillon and Dhillon (2003) suggested that since a significant quantity of Se is present in water-soluble form in alkaline soils, soil washing might be a potential technique for Se removal for such soils. Soil washing will not only remove Se from seleniferous soils, but Se can also be recovered during the subsequent treatment of soil leachate. Optimization of soil washing process parameters for treatment of artificially Se-contaminated soil has been investigated (Goh and Lim, 2004; Lim and Goh, 2005). However, there are no reports of washing of naturally occurring seleniferous soil. Loss of Se from soil due to rainwater and irrigation has been studied by Dhillon et al. (2008). Leaching and runoff of Se from a seleniferous soil from the northwest region of India showed that only around 0.29% Se was lost from the soil due to leaching and run off by rainfall (Dhillon et al., 2008).

In this research, soil washing is studied as a possible remediation strategy for removal and recovery of Se from seleniferous soil. Physico-chemical characteristics of the soil collected from the northwest region of India were studied. Soil column experiments were performed in order to understand the Se transport and migration pattern towards groundwater in case of rainfall and irrigation. The soil washing process was optimized by varying operational conditions and chemical nature of the soil washing solution to obtain the maximum Se removal. This study opens a new domain of seleniferous soil treatment and recovery of Se from Se-rich soils.

3.2. Materials and methods

3.2.1. Sample collection and characterisation

Soil samples were collected from the northwest region of India in the state of Punjab. The sample was taken at the geographical location 31° 07' 45.5"N and 76° 12' 43.1"E. The soil was collected from a sampling depth of 0-22 cm of this agricultural region. The soil was air-dried, sieved and stored at room temperature for further analysis. The physical properties such as particle size distribution, porosity, water retention capacity, bulk density and moisture content were determined by standard methods (USDA NRCS, 2014). The chemical properties of the soil such as pH, total organic carbon (TOC) and cation exchange capacity (CEC) in the soil were also determined according to USDA NRCS (2014). Heavy metals and elements were extracted from the soil by microwave-assisted acid digestion (EPA, 2001) using a MilestoneSTART D microwave oven as described by Ferraro et al. (2017). The total Se in soil digested was measured using atomic fluorescence spectrometry (AS60, AFS-8220, Fulltech Instrument, Roma, Italy). The acid digested samples were analysed for heavy metal contents using Inductively Coupled Plasma Mass Spectrometry (ICP-MS, XSeriesII, Thermo Scientific, Germany).

The sequential extraction for determination of the Se content in soil fractions was carried out employing two different methods (Martens and Suarez, 1997; Pueyo et al., 2008). Martens and Suarez (1997) used as a standard protocol for Se extraction from water soluble, adsorbed, organically associated and refractory fraction. The procedure was modified by an addition of a fifth step to completely digest the soil in aqua regia after step 4 and extract the residual Se. Pueyo et al. (2008) used a modified-BCR procedure for sequential extraction of heavy metals from soil. This method allowed Se extraction from different soil fractions, viz. exchangeable and weak acid soluble, reducible, oxidisable and residual fraction. The total Se in the soil and the cumulative Se fractions from both procedures were compared. All analytical-grade chemicals were purchased from Merck.

3.2.2. Soil flushing

In order to understand the transport of Se in soil, seleniferous soil column leaching tests were studied in duplicates. The experimental set-up was designed by stacking two 50 mL polypropylene centrifuge tubes above each other, where the top tube was used as a soil column while the bottom tube acted as a soil leachate collection unit. This set-up was designed as described by Hauser et al. (2005). The soil columns were supported by packing material comprising of glass wool and glass beads and 40 g soil was uniformly packed in the columns. Ultrapure water was pumped from the top of the column with the flow rate of 1 mL h^{-1} and allowed to flush out of the column by gravity. Liquid samples were collected at the bottom of the column at regular intervals, its volume measured and analysed for total Se and organic matter content. The experiment was stopped when no more Se was leached from the column. To analyze soluble, yet unleached Se in the soil column, the soil from each column was mixed with water under anaerobic conditions at room temperature at 150 rpm for 3 h. The aqueous phase of the leachate was separated by centrifugation at 5000 rpm for 15 min and analyzed for its Se content. The Se concentration in the liquid samples was determined by digestion with concentrated hydrochloric acid at 90 °C for 30 min, followed by suitable dilution with ultrapure water and measurement using AFS. The results from these tests were used to calculate the Se mass balance.

3.2.3. Soil washing

Description of tests carried out by varying different operational parameters (such as treatment time, multi-washing, temperature, pH, liquid to solid (L:S) ratio, competing ions and oxidizing agents) is provided in **Table 3.1**. After optimization of L:S ratio, treatment time and temperature, 1.5 g soil sample was mixed with 30 mL washing solution and incubated at room temperature on a shaker at 150 rpm for 3h for the remaining tests (pH, competing ions and oxidising agents). All experiments were conducted in triplicates. After soil washing tests, the samples were centrifuged as described for soil flushing experiments in order to separate treated soil samples from the soil leachates. The supernatant was analysed to determine Se concentration and organic matter. The organic matter content of the samples was analysed by acquiring three-dimensional excitation and emission matrix fluorescence (3DEEM) using a LS 45 spectrofluorimeter (Perkin Elmer, USA).

Table 3.1. Description of expriments and variation of operational parameters for optimisation of the seleniferous soil washing process

Parameters	Range	Description
Treatment time	3 h, 6 h, 1 d, 2 d, 3 d and 5 d	1.5 g soil in 30 mL ultrapure water mixed at 150 rpm at room temperature
Multi-washing	3 times	1.5 g soil in 30 mL ultrapure water mixed at 150 rpm at room temperature
Treatment temperature	20, 30, 40, 50 and 60 °C	1.5 g soil in 30 mL ultrapure water mixed at 150 rpm
Competing ions (SO_4^{2-} and PO_4^{3-})	1:1, 1:2, 1:5, 1:10, 1:20 and 1:50 (Se to competing ions molar ratios)	SO_4^{2-} and PO_4^{3-} as Na_2SO_4 and Na_3PO_4, respectively
L:S ratio	100:1, 100:1.5, 100:3, 100:5, 100:10, 100:20 (v/w)	With ultrapure water and competing ions (molar ratio 1:1)
pH	6, 7, 8, 9 and 10	pH of water was adjusted with 0.1 M HCl and 0.1 M NaOH
Oxidising agent ($KMnO_4$, H_2O_2 or $K_2S_2O_8$)	The soil was subject to the oxidants at different concentrations of 0.1, 0.25, 0.5, 1, 2 and 4%	v/v for H_2O_2 and w/v for $KMnO_4$ and $K_2S_2O_8$

3.3. Results

3.3.1. Soil characterisation

The pH of the seleniferous soil was 8.15 (±0.10), while its electrical conductivity was 326 (±8) $\mu S\ cm^{-1}$. Total organic carbon content, water holding capacity, and bulk density were 255 (±23) mg kg^{-1}, 750 (±50) mL kg^{-1}, and 1.204 (±0.024) kg L^{-1}, respectively. The initial moisture content of the soil was 1.95 (±0.15) % and its cation exchange capacity was 21.0 (±0.8) meq Na per 100 g soil. Particle size distribution analysis showed the sand:silt:clay content of the soil in the ratio of 4.1: 21.6: 74.3. According to the USDA (1987), the soil can be classified as clay. The total Se content in soil was 4.75 ($\pm$0.02) mg kg^{-1}. **Table 3.2** defines the total concentration of metals in soil resulting from ICP-MS analysis.

Table 3.2. Total elemental concentrations in the seleniferous soil

Element	Al	Ca	Cr	Cu	Fe	K	Mg	Mn	Na	Ni	Zn
Concentration (g kg^{-1})	4.20	9.90	0.02	0.02	10.03	1.06	2.99	0.18	1.82	0.01	0.09

Table 3.3. Se concentration in seleniferous soil fractions detected for each step of sequential extraction procedures

Extraction steps	Pueyo et al. (2008)		Martens et al. (1997)	
	Se fraction	Se (mg kg^{-1})	Se fraction	Se (mg kg^{-1})
Step 1	Exchangeable and weak acid soluble	1.19 $\pm$ 0.06	Water soluble Se	0.72 $\pm$ 0.03
Step 2	Reducible fraction	0.24 $\pm$ 0.03	Adsorbed Se	0.16 $\pm$ 0.01
Step 3	Oxidisable fraction	3.17 $\pm$ 0.33	Organically associated Se	2.17 $\pm$ 0.09
Step 4	Residual fraction	0.18 $\pm$ 0.02	Refractory Se	1.14 $\pm$ 0.04
Step 5			Residual Se	0.77 $\pm$ 0.15
Total Se		4.78 $\pm$ 0.43		4.96 $\pm$ 0.32

The results from the sequential extraction procedures of the Se concentration in soil fractions are reported in **Table 3.3.** According to the sequential extraction procedure described by Pueyo et al. (2008), Se was mainly extracted in 2 major fractions: the exchangeable and weak acid soluble (24.9%) and oxidisable (66.3%) fraction. Results from sequential extraction procedure described by Martens et al. (1997) displayed a more broad Se distribution among the various soil fractions. In this case, the highest Se content was present in the organically associated or oxidisable fraction (45.7%), followed by the refractory (24%), residual (16.2%) and water soluble (15.2%) fraction. The total Se extracted via both procedures (Martens and Suarez, 1997; Pueyo et al., 2008) corresponds to the total Se content(4.75 $\pm$0.02 mg kg^{-1}) measured in the soil during its characterisation.

3.3.2. Se leaching in column by soil flushing

Visual inspection showed that the soil structure remained intact, while the ultrapure water percolated uniformly from the column at 1 mL h^{-1} during the entire column leaching experiment. The total amount of Se leached from each column was measured with respect to the total volume of leachate collected (**Figure 3.1a**). The column consisting of 40 gram soil can be divided into 4 layers, each containing 10 g. The amount of removed Se decreased as the quantity of the soil layers increased. It was observed that 13.6, 10.4, 5.5 and 4.1% Se was extracted from 1, 2, 3 and 4 layers of soil, respectively.

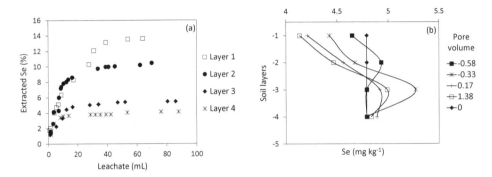

Figure 3.1. (a) Se extracted (%) from the soil with respect to the total leachate and (b) soluble Se migration across soil column layers with respect to different pore volume

The total soluble unleached Se in the columns at the end of the experiment was found to be below the detection limit, suggesting that no more soluble Se was present in the soil columns. Total Se concentration analysis in different soil layers after the extraction also revealed that Se was removed from the upper layers and was deposited in the lower layers during the course of the leaching procedure. However, there was neither depletion nor deposition of Se in the lower layers by the end of the experiment (**Figure 3.1b**).

3.3.3. Optimisation of soil washing

The results showed maximum Se extraction from soil of 23.94 ($\pm$0.33) % after 2 day of incubation (**Figure 3.2**). However, there was no significant difference in terms of amount of Se extracted when comparing the soil washing batches at all time intervals. Accordingly, 3h incubation time was selected as optimal value for the subsequent tests. Results from multiple washing test showed that extracted Se was around 8% and 3.5% in the second and third washing step respectively for all incubation times (**Figure 3.2**). In subsequent washing steps, the Se content in the soil leachates was less than the minimum detection limit of the AFS.

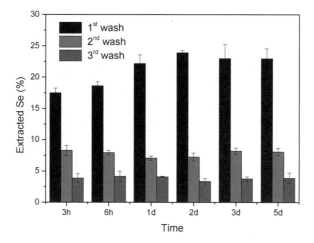

Figure 3.2. Se extracted from soil using ultrapure water at different incubation times and multi-washing steps

The effect of the incubation temperature on the Se extraction was studied by incubating the soil and water with a L:S ratio of 20:1 at 20, 30, 40, 50 and 60 °C. It was observed that the increase in temperature only slightly improves the Se extraction from the soil with no statistically significant differences in terms of extracted Se (**Figure 3.3**). Therefore, further experiments were performed at ambient temperature.

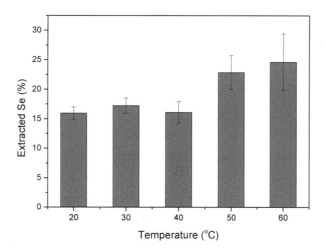

Figure 3.3. Se extracted from soil using ultrapure water at L:S ratio of 100:20 and different incubation temperatures

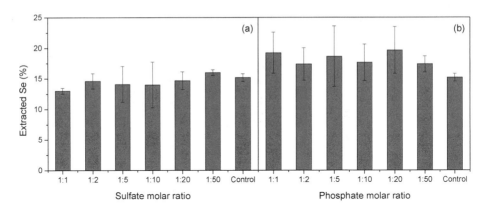

Figure 3.4. Se extracted in the presence of ultrapure water as control test and (a) sulfate and (b) phosphate as competing ions at different molar ratios at L:S ratio of 100:20

The effect of ultrapure water as control as well as SO_4^{2-} and PO_4^{3-} ions at different molar ratios (expressed as ratio between Se moles and competing ion moles) on the Se extraction from soil was studied (**Figure 3.4**). The overall higher extraction efficiency was observed with PO_4^{3-} ions compared to SO_4^{2-} ions. For instance, results displayed a Se extraction efficiency of 19.2 (±3.4) % for PO_4^{3-} ions as compared to 13.0 (±0.5) % for SO_4^{2-} ions for the 1:1 molar ratio (**Figure 3.4**). However, these results display no significant difference at different molar ratios for both SO_4^{2-} and PO_4^{3-} ions. Thus, the molar ratio of 1:1 for both competing ions was used to perform further experiments with different L:S ratio.

Effect of different L:S ratio involving both competing ions and ultrapure water as control was evaluated (**Figure 3.5**). These results show the maximum amount of Se (19.1 ± 0.3%) was extracted during the control test at a L:S ratio of 100:15 (v/w). However, the Se extracted was almost similar for all L:S ratios irrespective of the washing solution. The Se speciation in the soil largely depends on the soil pH and redox potential (Mayland et al., 1989). Therefore, the effect of variable pH and oxidising conditions was studied for optimisation of the seleniferous soil washing. Similar Se extraction efficiency was observed for all investigated pH values (**Figure 3.6**). Also, the final pH of the soil leachate for all batches was around 7.80 (±0.03). This was attributed to the high buffering capacity of the soil.

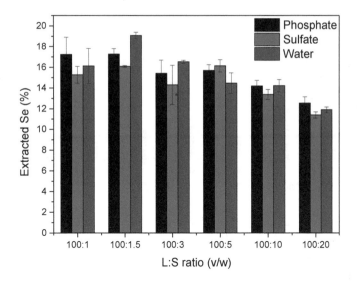

Figure 3.5. Se extracted in the presence of ultrapure water as control test and sulfate and phosphate as competing ions at different L:S ratios

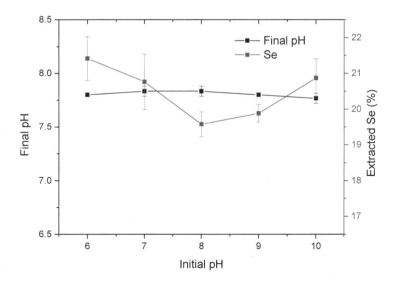

Figure 3.6. Se extracted from seleniferous soil at different pH and final pH values after the tests.

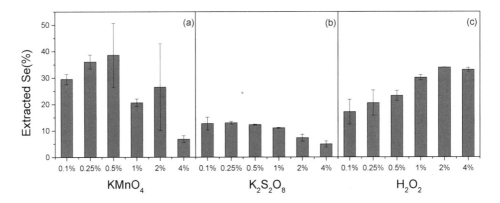

Figure 3.7. Effect of oxidising agents - $KMnO_4$ (a), $K_2S_2O_8$ (b) and H_2O_2 (c) on Se extraction from seleniferous soil

On the contrary, significant differences in terms of Se extraction efficiency were observed with the involvement of different oxidising agents. According to data obtained from tests at different concentrations of oxidising agents (**Figure 3.7**), the maximum Se extraction efficiency of 38.6 ($\pm$12.1) % was obtained in the presence of 0.5% $KMnO_4$, followed by 33.8 ($\pm$0.1) % in the presence of 2% H_2O_2. The extracts from both soil flushing and soil washing experiments were yellowish in colour due to the presence of organic matter in the effluent. 3DEEM analysis on effluent from both soil washing and flushing shows the presence of humic and fulvic substances as described by Chen et al. (2003). Based on the fluorescence signal, the organic matter content in the soil flushing extract was quantitatively much higher than that of the soil washing extract (**Figure 3.8**).

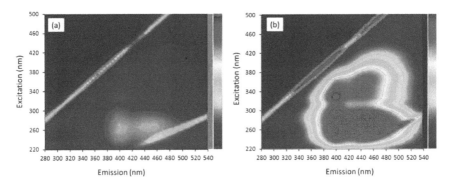

Figure 3.8. 3DEEM spectra of (a) soil leachate after washing seleniferous soil with ultrapure water in L:S ratio of 100:15 and (b) soil leachate from a 40 g soil flushing column after 72 h leaching

3.4. Discussion

3.4.1. Se migration in the soil column

In the regions of Punjab (India), the lithogenic origin of Se in the agricultural soil is from the rocks in the northern sub-Himalayan ranges (Shiwalik hill). Se in sediments is transported through seasonal rivulets and deposited in the affected regions of Punjab (Dhillon and Dhillon, 2014). A high Se content in the rock samples of the lower (1864-2754 $\mu g\ L^{-1}$) and upper (11-847 $\mu g\ L^{-1}$) Shiwalik ranges was observed (Dhillon and Dhillon, 2014).

Se migration across the soil column and its leaching into groundwater was studied by stimulating rainfall or irrigation in an *in situ* soil flushing set-up. In column experiments, the Se removal efficiency decreased with increase in column height. The migration pattern of Se (**Figure 3.1**) shows that the soluble Se from the upper layer migrates and settles in the lower layers of the soil columns. The absence of Se in the subsequent anaerobic washing of the soil in the column suggests that the soluble Se was reduced to unavailable forms during its migration through the soil column due to lack of aeration, thus accumulating in the lower layers of column. A similar trend was observed by Tokunaga et al. (1997), where reduction of Se from Se(VI) to Se(0) was observed in the depth profiles of the columns containing seleniferous soil from Kesterson reservoir.

Nevertheless, the insoluble Se fraction may be slowly reconverted back into soluble Se forms and contaminate groundwater with time. Bajaj et al. (2011) observed toxic concentrations of Se (45-341 $\mu g\ L^{-1}$) in the groundwater in certain areas in the region of Punjab (India). The authors suggested that the prevailing intensive irrigation practices in the region by Se-enriched groundwater use are responsible for the high Se-concentrations in soil. Lapworth et al. (2017) assessed the groundwater quality of the northwest region of India and observed high concentrations of Se (10-40 $\mu g\ L^{-1}$) in the shallow groundwater system of the region. This was attributed to the mobilization of Se from soil and its transport to the oxidising conditions in the shallow aquifers in the region. The authors also observed that high Se concentrations are not only associated with the oxidising conditions, but also high NO_3^- concentrations in groundwater.

3.4.2. Optimization of seleniferous soil washing

The washing of the seleniferous soil was optimized in order to achieve maximum removal of Se from soil in the form of Se oxyanions (selenate and selenite). The oxyanions are the water-soluble and bioavailable forms of Se that after leaching into a suitable liquid medium can be treated by conventional continuous bioreactor systems that have been widely investigated (Tan et al., 2016).

Based on results from sequential extraction on seleniferous soil, optimal extraction of the water soluble Se fraction was evaluated by varying the treatment time, temperature, L:S ratio and pH of ultrapure water. An increase in temperature has been reported to improve Se leaching from the soil (Dhillon and Dhillon, 2003). Accordingly, around 24% of Se removal was observed during soil washing tests at 60 °C (**Figure 3.3**). Previous study reported that alkaline environmental conditions are favourable for oxidation of Se (Seby et al., 1997). Nonetheless, tests in the present study showed that the alkaline pH of the extractant did not affect the Se extraction from the soil. The final pH of the Se leachate was around 8 for all effluents, irrespective of their initial pH (**Figure 3.6**). This can be attributed to the high buffering capacity of the soil (Bolan et al., 2014) that led to no significant Se removal variation due to change in pH of the washing solution. Multi-washing of the soil has been reported to positively influence heavy metal extraction of Cu, Cd and Zn (Gusiatin and Klimiuk, 2012). In contrast, in this study, multi-washing of seleniferous soil did not significantly affect the Se removal from the soil. Therefore, use of high temperature, higher incubation time and multi-washing of the soil did not significantly improve the Se extraction from the seleniferous soil investigated and might also be an economical constraint for large scale applications.

Literature studies report that ions such as SO_4^{2-} and PO_4^{3-} compete with SeO_4^{2-} and SeO_3^{2-} for adsorption sites on organic matter in soil (Lim and Goh, 2005; Wright, 1999). The addition of these competing ions as Na_2SO_4 and Na_3PO_4 to the washing agent can be a feasible solution to increase Se extraction from soil. In the present work, however, no significant difference was observed in terms of Se extraction (**Figure 3.4**). On the contrary, soil washing carried out with water as washing agent displayed a better Se extraction efficiency than tests with competing ions at different L:S ratio (**Figure 3.5**). This may be due to significantly lower concentrations of competing ions that were applied during soil washing in this study compared to previous studies (Lim and Goh, 2005).

Alkaline pH, high redox potential as well as the organic carbon and clay content of the seleniferous soil suggest the presence of Se oxyanions in the soil (Bajaj et al., 2011; Moreno et al., 2013; Schilling et al., 2015). Sequential extraction of the soil suggests Se is mainly present in the oxidisable and organically associated fraction (**Table 3.3**). Therefore, oxidising agents such as $KMnO_4$, $K_2S_2O_8$ and H_2O_2 were applied in order to recover the oxidisable Se fraction from the soil. Results showed optimal Se extraction for soil washing with $KMnO_4$ suggesting a positive influence on Se leaching through oxidising agents (**Figure 3.7**). However, use of oxidising agents on seleniferous soils may render further agricultural purposes after treatment useless and studies to determine the fertility of the soil after such treatment are required. Moreover, the oxidised Se species (selenate and selenite) are water soluble and therefore, more bioavailable (Hagarova et al., 2003). According to this, water was suggested as an optimal soil washing solution in this research.

3.4.3. Practical implications

Ex situ approaches for seleniferous soil treatment have been unexplored in comparison to *in situ* soil remediation, probably due to their cost factor, Se speciation and low Se concentration in soil. However, recovery of this scarce element (Nancharaiah and Lens, 2015) from soil may prove to be a promising approach towards meeting the ever-increasing industrial demand of Se. One of the common *ex situ* approaches applied for remediation of metal-contaminated soils is soil washing for extraction of metal contaminants and treatment of washing solutions (Jiang et al., 2011; Kim et al., 2012). Soil washing is considered as a pre-concentration stage prior to further treatment of the leachate (Dermont et al., 2008). Several examples of soil washing technique application are reported for removal of metals like arsenic, cadmium, chromium, copper, mercury, nickel, lead and zinc from contaminated soils in Europe (Dermont et al., 2008; Kim et al., 2012; Mohanty and Mahindrakar, 2011, Satyro et al., 2017).

In case of soil washing in batch bottles, constant shaking in the incubator provides sufficient aeration which leads to complete removal of soluble Se. Also the different contact mode between the contaminated soil and washing solution should be taken into account, since it highly differs between the soil washing and flushing processes. It could also influence the Se extraction/leaching efficiency from the contaminated soil. The major disadvantage of soil flushing on seleniferous soils is that the height of the soil column negatively affects the Se removal efficiency. However, on a large scale, heap leaching could be carried out by percolating water through a shallow bed of seleniferous soil in order to maximize the Se removal. Nonetheless, constant mechanical pumping of the leaching solution into the soil column may increase the cost of soil treatment along with process costs due to soil excavation. Similarly, in case of soil washing in batches, although 20% of the Se is removed from the soil, constant mechanical stirring and excavation costs can increase the overall process costs.

Se extraction from soil washing can be combined with aerobic or anaerobic bioreactors (Espinosa-Ortiz et al., 2014; Tan et al., 2016) for treatment of leachate and possible recovery of Se in the form of elemental Se. This approach can also be explored for the treatment of Se contaminated soils with a possibility to recover this valuable pollutant for use in technological applications (Mal et al., 2016).

3.5. Conclusion

In this research, soil flushing was performed to assess Se migration in a soil column towards groundwater in case of rainfall and irrigation. Se migration and accumulation to the lower layers in the soil columns was observed suggesting reduction of soluble Se to insoluble Se forms. Nevertheless, with time, the insoluble Se fraction may be slowly re-oxidised to soluble Se forms and contaminate groundwater. Furthermore, washing of seleniferous soil was optimized in order to remove soluble and bioavailable Se species from soil. Among the various parameters investigated (treatment time, temperature, pH, L:S ratio, competing ions and oxidizing agents) for seleniferous soil, only oxidising agents showed the promising outcome with maximum Se removal. However, further research is required to check the feasibility and the cost of entire process including soil excavation, washing, effluent treatment followed by soil restoration to the environment for further agricultural use.

References

Ashworth, D.J., Shaw, G., 2006. Soil migration, plant uptake and volatilisation of radio-selenium from a contaminated water table. Sci. Total Environ. 370(2-3), 506–514.

Bajaj, M., Eiche, E., Neumann, T., Winter, J., Gallert, C., 2011. Hazardous concentrations of selenium in soil and groundwater in North-West India. J. Hazard. Mater. 189(3), 640–646.

Bolan, N., Kunhikrishnan, A., Thangarajan, R., Kumpiene, J., Park, J., Makino, T., Kirkham, M.B., Scheckel, K., 2014. Remediation of heavy metal(loid)s contaminated soils - To mobilize or to immobilize? J. Hazard. Mater. 266, 141–166.

Chen, W., Westerhoff, P., Leenheer, J.A, Booksh, K., 2003. Fluorescence excitation - Emission matrix regional integration to quantify spectra for dissolved organic matter. Environ. Sci. Technol. 37(24), 5701–5710.

Dermont, G., Bergeron, M., Mercier, G., Richer-Laflèche, M., 2008. Soil washing for metal removal: A review of physical/chemical technologies and field applications. J. Hazard. Mater. 152(1), 1–31.

Dhillon, K.S., Dhillon, S.K., 2014. Development and mapping of seleniferous soils in northwestern India. Chemosphere 99, 56–63.

Dhillon, K.S., Dhillon, S.K., 2003. Distribution and management of seleniferous soils, in: Sparks, D.L. (Ed.), Advances in Agronomy. Vol. 79, pp. 119–184.

Dhillon, S.K., Dhillon, K.S., Kohli, A., Khera, K.L., 2008. Evaluation of leaching and runoff losses of selenium from seleniferous soils through simulated rainfall. J. Plant Nutr. Soil Sci. 171, 187–192.

El-Ramady, H., Abdalla, N., Alshaal, T., Domokos-Szabolcsy, É., Elhawat, N., Prokisch, J., Sztrik, A., Fári, M., El-Marsafawy, S., Shams, M.S., 2015. Selenium in soils under climate change, implication for human health. Environ. Chem. Lett. 13(1), 1-19.

EPA, 2001. Method 3051, Microwave Assisted Digestion of Sediments, Sludges, Soils and Oils. Environmental Protection Agency, USA.

Espinosa-Ortiz, E.J., Gonzalez-Gil, G., Saikaly, P.E., van Hullebusch, E.D., Lens, P.N.L., 2015. Effects of selenium oxyanions on the white-rot fungus *Phanerochaete chrysosporium*. Appl. Microbiol. Biotechnol. 99(5), 2405–2418.

Ferraro, A., Fabbricino, M., van Hullebusch, E.D., Esposito, G., 2017. Investigation of different ethylenediamine-N,N'-disuccinic acid-enhanced washing configurations for remediation of a Cu-contaminated soil: process kinetics and efficiency comparison between single-stage and multi-stage configurations. Environ. Sci. Pollut. Res. 24(27), 21960–21972.

Goh, K.-H., Lim, T.-T., 2004. Geochemistry of inorganic arsenic and selenium in a tropical soil: effect of reaction time, pH, and competitive anions on arsenic and selenium adsorption. Chemosphere 55(6), 849–859.

Gupta, M., Gupta, S., 2017. An overview of selenium uptake, metabolism, and toxicity in plants. Front. Plant Sci. 7, 2074.

Gusiatin, Z.M., Klimiuk, E., 2012. Metal (Cu, Cd and Zn) removal and stabilization during multiple soil washing by saponin. Chemosphere 86(4), 383–391.

Hagarova, I., Zemberyova, M., Bajcan, D., 2003. Sequential and single step extraction procedures for fractionation of selenium in soil samples. Chem. Pap. 59(2), 93–98.

Hauser, L., Tandy, S., Schulin, R., Nowack, B., 2005. Column extraction of heavy metals from soils using the biodegradable chelating agent EDDS. Environ. Sci. Technol. 39, 6819–6824.

Jiang, W., Tao, T., Liao, Z., 2011. Removal of heavy metal from contaminated soil with chelating agents. Open J. Soil Sci. 1, 70–76.

Kim, K., Cheong, J., Kang, W., Chae, H., Chang, C., 2012. Field study on application of soil washing system to arsenic- contaminated site adjacent to J . refinery in Korea, in: International Conference on Environmental Science and Technology. IACSIT Press, Singapore, pp. 1–5.

Kirk, L.M.B., 2014. *In situ* microbial reduction of selenate in backfilled phosphate mine waste, S.E. Idaho. [Doctoral thesis]. Montana State University, Bozeman, Montana, USA.

Lapworth, D.J., Krishan, G., MacDonald, A.M., Rao, M.S., 2017. Groundwater quality in the alluvial aquifer system of northwest India: New evidence of the extent of anthropogenic and geogenic contamination. Sci. Total Environ. 599–600, 1433–1444.

Lenz, M., Lens, P.N.L., 2009. The essential toxin: the changing perception of selenium in environmental sciences. Sci. Total Environ. 407(12), 3620–3633.

Lim, T.-T., Goh, K.-H., 2005. Selenium extractability from a contaminated fine soil fraction: implication on soil cleanup. Chemosphere 58(1), 91–101.

Mackowiak, C.L., Amacher, M.C., 2007. Soil sulfur amendments suppress selenium uptake by alfalfa and western wheatgrass. J. Environ. Qual. 37(2), 772–729.

Mal, J., Nancharaiah, Y.V, van Hullebusch, E.D., Lens, P.N.L., 2016. Metal chalcogenide quantum dots: biotechnological synthesis and applications. RSC Adv. 6, 41477–41495.

Martens, D.A., Suarez, D.L., 1997. Selenium speciation of soil/ sediment determined with sequential extractions and hydride generation atomic absorption spectrophotometry. Environ. Sci. Technol. 31, 133–139.

Mayland, H.F., James, L.F., Panter, K.E., Sonderegger, J.L., 1989. Selenium in seleniferous environments, in: Jacobs, I. W., (Ed.), Selenium in agriculture and the environment. Am. Soc. Agron., Madison, pp.15–50.

Mohanty, B., Mahindrakar, A.B., 2011. Removal of heavy metal by screening followed by soil washing from contaminated soil. Int. J. Technol. Eng. Syst. 2(3), 290–293.

Moreno, R.G., Burdock, R., Cruz, M., Álvarez, D., Crawford, J.W., 2013. Managing the selenium content in soils in semiarid environments through the recycling of organic matter. Appl. Environ. Soil Sci. 2013, 283468.

Nancharaiah, Y.V, Lens, P.N.L., 2015. Selenium biomineralization for biotechnological applications. Trends Biotechnol. 33(6), 323–330.

Nancharaiah, Y.V, Mohan, S.V., Lens, P.N.L., 2016. Biological and bioelectrochemical recovery of critical and scarce metals. Trends Biotechnol. 34(2), 137–155.

Pueyo, M., Mateu, J., Rigol, A., Vidal, M., López-Sánchez, J.F., Rauret, G., 2008. Use of the modified BCR three-step sequential extraction procedure for the study of trace element dynamics in contaminated soils. Environ. Pollut. 152(2), 330–341.

Satyro, S., Race, M., Marotta, R., Dezotti, M., Guida, M., & Clarizia, L. (2017). Photocatalytic processes assisted by artificial solar light for soil washing effluent treatment. Environ Sci Pollut Res Int. 24(7), 6353-6360.

Schiavon, M., Pilon-Smits, E.A.H., 2017. The fascinating facets of plant selenium accumulation – biochemistry, physiology, evolution and ecology. New Phytol. 213(4), 1582–1596.

Schilling, K., Johnson, T.M., Dhillon, K.S., Mason, P.R.D., 2015. Fate of selenium in soils at a seleniferous site recorded by high precision Se isotope measurements. Environ. Sci. Technol. 49(16), 9690–9698.

Seby, F., Gautier, M.P., Lespks, G., Astruc, M., 1997. Selenium speciation in soils after alkaline extraction. Sci. Total Environ. 207(2-3), 81–90.

Sharma, S., Bansal, A., Dogra, R., Dhillon, S.K., Dhillon, K.S., 2011. Effect of organic amendments on uptake of selenium and biochemical grain composition of wheat and rape grown on seleniferous soils in northwestern India. J. Plant Nutr. Soil Sci. 174, 269–275.

Tan, L.C., Nancharaiah, Y.V, van Hullebusch, E.D., Lens, P.N.L., 2016. Selenium: Environmental significance, pollution, and biological treatment technologies. Biotechnol. Adv. 34(5), 886–907.

Tinggi, U., 2003. Essentiality and toxicity of selenium and its status in Australia : a review. Toxicol. Lett. 137(1-2), 103–110.

Tokunaga, T.K., Brown, G.E., Pickering, I.J., Sutton, S.R., Bajt, S., 1997. Selenium redox reactions and transport between ponded waters and sediments. Environ. Sci. Technol. 31(5), 1419–1425.

USDA NRCS, 2014. Soil Survey Field and Laboratory Methods Manual, Soil Survey Investigations. Lincoln, Nebraska.

USDA Textural Soil Classification, 1987. Soil Mechanics Level 1. Module 3. Study Guide. Soil Conservation Service, United State Department of Agriculture.

Wadgaonkar, S.L., Nancharaiah, Y.V., Esposito, G., Lens, P.N.L. Environmental impact and bioremediation of seleniferous soils and sediments. Crit. Rev. Biotechnol. *In press.*

Wright, W.G., 1999. Oxidation and mobilization of selenium by nitrate in irrigation drainage. J. Environ. Qual. 28(4), 1182-1187.

Wu, L., 2004. Review of 15 years of research on ecotoxicology and remediation of land contaminated by agricultural drainage sediment rich in selenium. Ecotoxicol. Environ. Saf. 57(3), 257–269.

CHAPTER 4

In situ and *ex situ* bioremediation approaches for removal and recovery of selenium from seleniferous soils of Northern India

This chapter has been modified and published as:

Wadgaonkar SL, Ferraro A, Nancharaiah YV, Dhillon KS, Fabbricino M, Esposito G, Lens PNL. (2018) *In situ* and *ex situ* bioremediation of seleniferous soils from Northwestern India. J. Soils Sediments. DOI: 10.1007/s11368-018-2055-7

Abstract

In this study, *in situ* and *ex situ* approaches were evaluated for bioremediation of seleniferous soils. As part of *in situ* remediation approach, effect of organic amendment and bioaugmentation was determined. The effect of soil amendment with different organic sources (fermentable, non-fermentable and non-hydrolysable electron donors) and bioaugmentation with anaerobic granular sludge was evaluated in soil microcosms for *in situ* reduction of Se oxyanions. Results showed no significant difference in the Se reduction profiles with or without organic amendment or bioaugmentation. This suggested that the indigenous Se reducing microorganisms and oxidisable organic carbon present in the soil are sufficient for *in situ* soil bioremediation. Moreover, *ex situ* recovery of selenium from the seleniferous soil was attempted. *Ex situ* approach involved evaluation of seleniferous soil leaching and Se-leachate treatment in an upflow anaerobic sludge blanket (UASB) reactor. In this *ex situ* approach, the seleniferous soil was leached and the resulting leachate was biologically treated in an UASB reactor with varying conditions of organic carbon supplementation. This approach has efficiently (90%) removed Se from the soil leachate and recovered from the anaerobic granular sludge. The treated Se-leachate leaving the UASB reactor contained Se at <5 μg L^{-1} which is in accordance with the USEPA guideline for the selenium discharge in wastewater. This showed the soil leachate biological treatment suitability for both significant Se recovery and environmentally sustainable effluent discharge.

Key words: Selenium bioreduction, soil washing biotreatment, UASB reactor, anaerobic granular sludge, organic amendment, bioaugmentation.

4.1. Introduction

Selenium (Se) occurs in four primary oxidation states in soil, viz. selenate [Se(VI)], selenite [Se(IV)], elemental selenium [Se(0)] and selenide [Se(-II)], mainly depending on the prevailing pH, redox potential and organic content of the soil (Hagarova et al., 2003). The common range of Se concentration in soils is 0.01–2 mg kg^{-1}, although its distribution variability can also range from almost zero to 1250 mg kg^{-1} (Oldfield, 2002). The Se content in soil is the result of either weathering of parent rocks across the Earth's crust or anthropogenic sources such as agricultural and industrial waste disposal (El-Ramady et al., 2015; Kabata-pendias and Pendias, 2001). By definition, soil containing less than 0.5 mg Se kg^{-1} soil are considered Se deficient soil, while those containing more than 5 mg Se kg^{-1} soil are seleniferous soil (Oldfield, 2002). However, the information on Se concentration in soil is not representative of its bioavailability. In fact, Se bioavailability also depends on Se speciation and Se complexation with organic and inorganic components of the soil (Frankenberger et al., 2004).

Se directly enters the food chain from the soil due to take up via or absorption onto plant roots and is then transferred across trophic levels (Ashworth and Shaw, 2006; De La Riva et al., 2014). Therefore, malfunctions associated with Se deficiency (Wang et al., 2016) and toxicity (Misra et al., 2015) are directly associated with the Se concentration and speciation in the soil. In case of Se deficient soil, a wide range of research is being carried out in order to optimize Se amendment (Moreno et al., 2013; Yasin et al., 2014) to the agricultural soil aimed at achieving the recommended dietary Se requirement for humans and animals (WHO 2011). In case of seleniferous soil, *in situ* bioremediation approaches such as phytoremediation (Lindblom et al., 2014; Yasin et al., 2014), organic amendment (Flury et al., 1997; Williams et al., 2013) and biovolatilization (Dhillon et al., 2010; Gadd, 2000) have been investigated. These techniques are aimed at either removing Se from the soil or converting the bioavailable Se forms into insoluble or volatilized forms.

In situ bioremediation of seleniferous soils through phytoremediation, soil amendment with organic matter and biovolatilization are considerably slow processes. This can amplify the risk of groundwater contamination by leaching and surface water contamination by deposition and solubilization of volatile forms. In a previous study, seleniferous soil collected from the state of Punjab (India) was characterized and a soil washing process for its remediation was optimized by varying physical and chemical parameters (Wadgaonkar et al., submitted). Soil washing is considered as one of the permanent and cost-effective techniques for the clean-up of contaminated soils and sediments (Dermont et al., 2008). It can be considered as a pre-concentration step where the contaminant is extracted into a small volume of soil leachate that can be further treated.

In this study, *in situ* bioremediation treatment of seleniferous soil was investigated though biostimulation by adding different electron donors as well as through bioaugmentation by adding anaerobic granular sludge to microcosms containing the seleniferous soil. Besides, *ex situ* bioremediation of soil washing leachate containing the water-soluble Se fraction was carried out in a lab-scale upflow anaerobic sludge blanket (UASB) reactor. The UASB reactor was operated at varying conditions using lactate (electron donor) dosing and the operational conditions were optimized for the removal and recovery of Se from the soil leachate.

4.2. Materials and methods

4.2.1. Sample collection and chemicals

Seleniferous soil was collected from an agricultural region of Northwest India in the state of Punjab. The geographical location of the seleniferous soil sample collection site is 31° 07' 45.5"N and 76° 12' 43.1"E. The soil was collected from a sampling depth of 0-22 cm. The soil was characterized in the previous study (Wadgaonkar et al., submitted). The pH of the soil was 8.15 ($\pm$0.10) and its total organic content was 255 ($\pm$23) mg kg^{-1}. Anaerobic granular sludge was collected from a paper mill wastewater treatment plant (Eerbeek, the Netherlands) and its characteristics are detailed in Roest et al. (2005). All analytical-grade chemicals were purchased from Merck.

4.2.2. *In situ* treatment using microcosms

Effect of bioaugmentation with anaerobic granular sludge
Bioaugmentation of the soil was done in microcosms by adding anaerobic granular sludge, suspended in a mineral salt medium containing essential salts and trace elements as described by Stams et al. (1992). 1.12 g L^{-1} sodium lactate was provided as the carbon source and electron donor. The pH of the medium was adjusted to 7 using 1 M HCl. The microcosms were prepared by incubating a measured quantity of soil in minimal salt medium containing electron donor at a liquid to solid (L:S) ratio of 10:1 (v/w). Microcosm setups were incubated under anaerobic conditions with 5% inoculum at 30 °C for 4 d. The serum bottles were closed with butyl rubber stoppers, crimp sealed and purged with N$_2$ for 5 min. The experiment was conducted in

triplicates. Liquid samples were collected after every 24 h and analysed to determine the extracted Se from the soil and the residual lactate concentration. The effect of indigenous soil microorganisms on seleniferous soil bioremediation was determined by setting up microcosms containing only 10% soil and 10 mM lactate. The role of only granular sludge on Se contaminated soil bioremediation was studied by adding heat sterilised soil and 5% granular sludge inoculum to the microcosm. A control was setup by adding heat-sterilized soil to the mineral salts medium.

Effect of biostimulation with different electron donors

Biostimulation through organic amendment for Se removal from the seleniferous soil was carried out by incubating soil in minimal salt medium containing different electron donors, i.e. 1.12 g L^{-1} lactate, 1.8 g L^{-1} glucose and 1 g L^{-1} maize husk. All electron donors were provided individually in excess in order to ensure complete Se reduction and incubated with a L:S ratio of 10:1 (v/w). Control setup contained only medium and soil was also incubated in order to determine the effect of the soil organic carbon content on Se removal. The incubations were carried out under anaerobic conditions at 30 °C for 7 d. Samples were collected every 24 h and analysed to determine total Se and acetate concentrations in the microcosms.

4.2.3. Seleniferous soil leachate preparation and UASB reactor operation

Soil washing tests based on soil washing optimization carried out in the previous study (Wadgaonkar et al., submitted) were performed by mixing seleniferous soil and demineralized water at a L:S ratio of 20:1 (v/w). Mixing was carried out overnight at room temperature (25 ± 2 °C) and an agitation speed of 180 rpm. The leachate was then separated from the washed soil by centrifugation at 4550 g for 15 min. The leachate was mixed with mineral salt medium as previously described for the microcosm bioaugmentation experiments. Sodium lactate (1 mM) was added as carbon source and electron donor to attain an organic loading rate (OLR) of 96 mg chemical oxygen demand (COD) L^{-1} day^{-1}. Considering the soluble Se concentration was around 55 μg L^{-1} in the soil leachate, COD was provided in excess based on the stoichiometric calculations at pH 7 where 1 mole lactate is required to reduce 2 moles of selenate to elemental Se (**Eq. 4.1**).

$$CH_3CH(OH)COO^-_{(aq)} + 2\ SeO_4^{2-}_{(aq)} + 2\ H^+_{(aq)} \leftrightarrows 3\ HCO_3^-_{(aq)} + 2\ Se^0_{(s)} + 2\ H_2O_{(aq)} \qquad \text{(Eq.4.1)}$$

A transparent polyvinyl cylinder with a working volume of 1.2 L was used as UASB reactor. The reactor was operated at ambient temperature (25 ± 2 °C), pH 7 and a hydraulic retention time (HRT) of 24 h. The biomass was suspended in the bioreactor at an upflow velocity of 0.2 m h^{-1} by recirculation. 120 mL (10% of the reactor working volume) of anaerobic granular sludge was added as inoculum.

The UASB reactor was operated in 5 phases with a gradual organic loading decrease in each phase. The organic loading decrease was aimed at determining the minimal organic requirement for complete Se removal from the seleniferous soil leachate. Phase I (day 1-19) was considered an acclimatization phase, with the addition of the mineral salt medium (Stams et al., 1992) to the soil leachate and pH adjusted to 7 using 0.1 M HCl. In addition, 1 mM sodium lactate was added as electron donor. In phase II (day 20-35), the reactor operation was stopped for maintenance work due to breakdown of a peristaltic pump. The granular sludge was collected from the reactor and stored at 4 °C until the reactor was restarted in phase III (day 36-46). Phase III represented a recovery phase and the reactor was maintained in similar conditions as phase I to achieve acclimatization of the microorganisms to recover from the former operational interruption. In phase IV (day 47-64), the UASB reactor was operated in feast and famine conditions, wherein the electron donor was provided intermittently after every 4 days. In this phase, the soil leachate was provided to the reactor without addition of mineral salt medium and the pH was not adjusted. In phase V (day 65-78), only soil leachate was used as the influent without adding electron donor and altering its pH or salt concentration.

4.2.4. Characterisation of anaerobic granular sludge

Anaerobic granule samples were collected from the UASB reactor at days 0, 45 and 78 of the reactor operation. The granules were washed with MQ water and dried at 105 °C for 2 hours. 1 g dried granules was digested with 9 mL of 63% concentrated nitric acid and 1 mL of 30% hydrogen peroxide in a microwave (Milestone S.r.l. - START D - Microwave Digestion System, Sorisole (BG) - Italy). Liquid samples from acid digestion were then filtered and analysed using atomic fluorescence spectroscopy to determine the total Se concentration in the granules.

The structure and composition of the granule samples were determined using SEM-EDS (Tescan Vega 3, Gambetti Binasco, Milan, Italy) and P-XRD (Rigaku Miniflex, Neu-Isenburg, Germany) analysis. For SEM-EDS and P-XRD analysis, the washed granules were freeze-dried in a lyophilizer (Christ Alpha 2-4 LSCplus, Osterode am Harz, Germany). For P-XRD spectra measurements, cobalt was used as X-ray source ($\lambda = 1.79$ Å) and the analysis was performed as described by Cennamo et al. (2016).

4.2.5. Analysis

Liquid samples were regularly collected from the *in situ* microcosms, soil washing influent and effluent of the UASB reactor and filtered through 0.45 μm syringe filters for further analysis. Se concentration was analysed after sample preparation through acid digestion carried out with concentrated hydrochloric acid. The samples were mixed with 37% HCl in the ratio of 1:1(v/v) and digested at 90 °C for 30 min according to the instrument manufacturer instructions. The digested samples were then suitably diluted with MQ water for analysis using atomic fluorescence spectroscopy (AFS-8220) coupled to an AS-60 autosampler (Titan, Fulltech Instruments, Roma, Italy). The lactate and acetate concentrations were measured by high pressure liquid chromatography (HPLC, LC 25 chromatography oven, Dionex) equipped with a 250 x 4.60 mm column (Synergi 4u Hydro RP 80A) and a UV detector (AD25 absorbance detector, Dionex) as described by Luongo et al. (2017). COD was measured using standard methods (APHA/AWWA/WEF, 2012). The three-dimensional excitation and emission matrix fluorescence (3DEEM) for the influent and the effluent from the UASB reactor were acquired through a LS 45 spectrofluorimeter (Perkin Elmer, USA) using the following excitation-emission range: ex. 220-500 nm and em. 280-540 nm (Pontoni et al., 2016).

4.3. Results

4.3.1. *In situ* microcosm for anaerobic reduction of selenium

Effect of bioaugmentation

In the microcosms bioaugmented with anaerobic granular sludge (**Figure 4.1a**), complete removal of dissolved Se from the medium was observed after 3 days of incubation, while complete lactate removal was achieved after 2 days. Similar results for Se and lactate removal were observed for microcosms containing only soil (**Figure 4.1b**) and those containing sterile soil with anaerobic granular sludge (**Figure 4.1c**). No Se and lactate removal was observed in control bottles (**Figure 4.1d**), suggesting that removal of soluble Se from the liquid medium was solely due to biological process. For all the microcosms, the pH of the medium was constant at 7.2 ($\pm$ 0.1) throughout the incubation period. The indigenous microorganisms showed a promising effect on Se removal. Therefore, subsequent biostimulation experiments with organic amendments were carried out using only the indigenous soil population for Se reduction and removal without bioaugmentation with granular sludge.

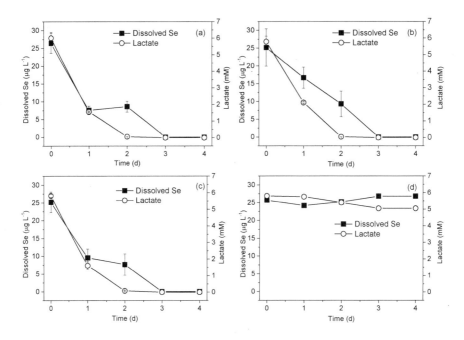

Figure 4.1. Effect of bioaugmentation on seleniferous soil bioremediation through involvement of (a) anaerobic granular sludge, (b) only soil with indigenous microbes, (c) sterile soil with anaerobic granular sludge and (d) control with sterile soil only

Effect of biostimulation with organic amendment

Controls without organic amendment (**Figure 4.2a**) were compared with microcosms amended with lactate, glucose and maize husk (**Figures 4.2b, c and d**). Significant Se removal was observed after one day of incubation in microcosms containing lactate while after 2 days for the tests with glucose and the control. Se removal occurred after 3 days of incubation when maize husk was used as electron donor. After 7 days of incubation, acetate concentrations were 7.70 ($\pm$ 0.35), 66.97 ($\pm$ 1.25), 236.02 ($\pm$ 6.21) and 1911.70 ($\pm$ 89.10) mg L^{-1} in the control microcosms and those with lactate, glucose and maize husk as electron donor, respectively (**Figure 4.2**).

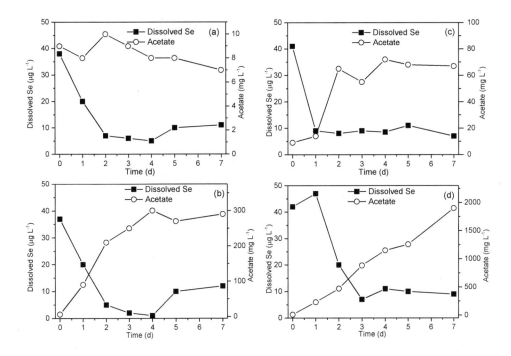

Figure 4.2. Effect of biostimulation with organic amendment on Se and acetate concentrations during seleniferous soil bioremediation in (a) control test and experiments with (b) lactate, (c) glucose and (d) maize husk as electron donor

Changes in the pH for the microcosms containing glucose and maize husk were noticeable, while pH was almost constant and equal to 7 for the control and lactate microcosms (**Figure 4.3**). A significant drop in pH values from 7 to 5 was observed in the microcosms amended with glucose and maize husk after 24 h of incubation. On day 3, pH increase up to almost 6 was observed in the microcosms amended with glucose. Subsequently, pH suddenly dropped to 5.5 on day 4. In case of maize husk microcosms, constant pH values of 5 were observed for the remainder of the incubation period.

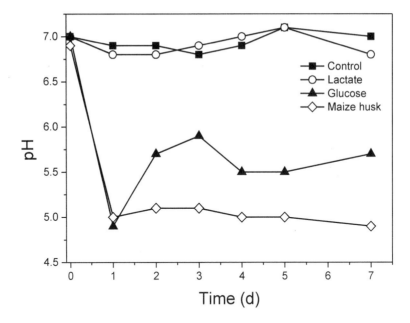

Figure 4.3. Evolution of the pH for control microcosms and microcosms amended with lactate, glucose or maize husk

4.3.2. Treatment of soil leachate in a UASB reactor

pH profile

Figure 4.4a shows the pH trend for the influent and effluent during the different phases of the UASB reactor run. In phases I and III, the influent pH was adjusted to 7 in order to provide optimal conditions for Se reduction by the anaerobic granular sludge. In contrast, in phases IV and V, the Se leachate was fed directly to the UASB reactor without pH adjustment. The influent pH in phases IV and V rose to around 8, corresponding to the natural alkaline pH of the seleniferous soil involved in the leaching test to extract the water soluble Se fraction. Nevertheless, the effluent pH of the UASB reactor was also almost constant at pH 7 during all the phases.

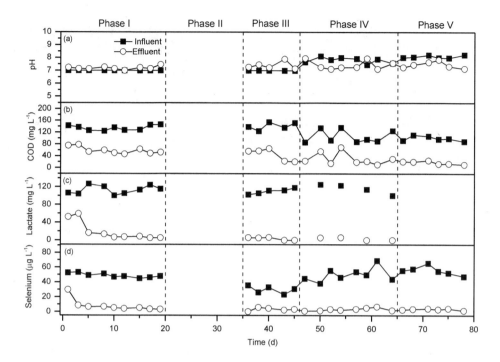

Figure 4.4. Evolution of (a) pH, (b) COD, (c) lactate and (d) Se concentrations of the influent and effluent of the UASB reactor treating seleniferous soil leachate during different phases

COD and lactate removal

The COD and lactate removal profiles are illustrated in **Figures 4.4b and 4.4c**, respectively. For phases I and III, significant COD concentration was detected in the UASB effluent due to the presence of high concentrations of humic and fulvic substances as seen in the 3DEEM spectra (**Figure 4.5**) extracted from soil during the leaching treatment. The COD removal increased in phases IV and V. However, a residual COD concentration of around 40 mg L^{-1} was always present in the treated leachate. The residual COD may correspond to non-biodegradable organic matter possibly deriving from the EPS and humic substances produced by the microbial community in the anaerobic granular sludge (Cassidy et al., 2017). The lactate concentration during the first 3 days (Phase I) of the reactor run was around 50 mg L^{-1}, after which complete lactate removal was observed in the subsequent phases (**Figure 4.4c**).

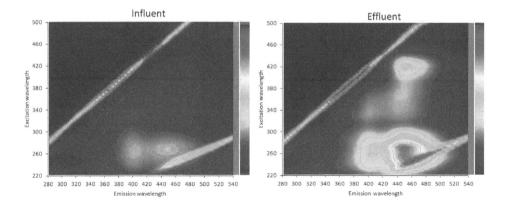

Figure 4.5. 3DEEM image of the Se-leachate before (influent) and after (effluent) treatment by the UASB reactor (Phase V)

Selenium removal

A low Se removal efficiency (**Figure 4.4d**) was observed at the start of the experiment with a value around 44% on day 1. However, the Se removal efficiency quickly increased to 90% during the subsequent measurements achieving an effluent Se concentration lower than 5 µg L^{-1}. Also, results showed that reactor operation interruption (phase II), feast and famine conditions (phase IV) and complete omission of lactate from the reactor influent (phase V) did not affect the Se removal efficiency.

4.3.3. Characterisation of anaerobic granular sludge from the UASB reactor

The total Se concentration in the anaerobic granular sludge amounted to 43.5 ($\pm$ 0.7) µg Se per gram of granular sludge. SEM analysis shows significant variations in the individual granular sludge surface during the course of the reactor run (**Figure 4.6**). The EDS analysis provided evidence of Se deposition on the surface of the granular sludge and confirmed the chemical nature of the crystal deposits to be Se(0). The elemental composition (atom %) of the selenite reducing granules was: Ca 32.2, O 25.6, C 21.3, N 3.8, Se 2.5, Fe 0.8. The signals of elements like carbon, oxygen, nitrogen and sulfur are emanating from the granular sludge and represent the microbial cells and the EPS matrix. Other inorganic elements may originate from the soil leachate extracted during the soil washing procedure. The P-XRD analysis (**Figure 4.7**) showed a good agreement with the Se peaks when compared with the reference spectra from the standard database (Se-JCPDS No. 03-065-3404). The maximum intensity was observed for Se at 34.34° (101), which corresponds to the hexagonal lattice structure of Se.

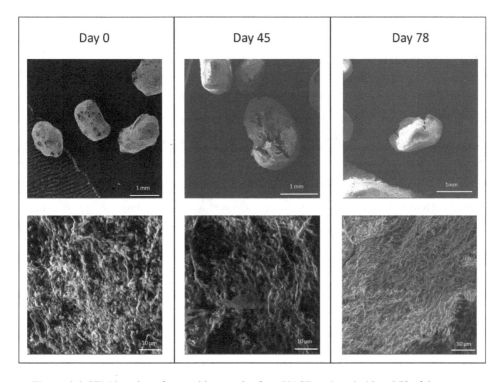

Figure 4.6. SEM imaging of anaerobic granules from UASB at days 0, 45 and 78 of the reactor operation

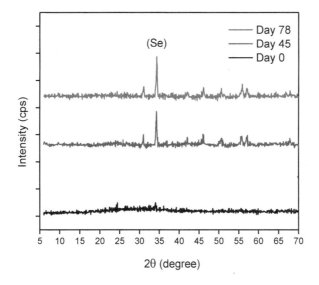

Figure 4.7. P-XRD analysis for the granular sludge at days 0, 45 and 78 of the UASB reactor operation

4.4. Discussion

4.4.1. *In situ* biotreatment of selenium oxyanions

Biological treatment of seleniferous soil shows that indigenous soil microorganisms and the organic content present in the soil are sufficient for Se removal under *in situ* conditions (**Figures 4.1 and 4.2**). Similar results were obtained in a field experiment that assessed microbial volatilization of Se as a bioremediation approach to dissipate Se at the Kesterson reservoir (California, USA) (Flury et al. (1997). To enhance microbial methylation of Se to volatile $(CH_3)_2Se$, the plots containing Se-contaminated agricultural drainage were augmented with different carbon (cattle manure, gluten, orange peel and straw of *Typha latifolia*) and protein (casein) sources and were periodically tilled and irrigated. Soil amendment with different organic and protein sources was hypothesized to improve microbial activity and stimulate bioremediation. The authors observed that the highest amount of Se depletion occurred with amendment of the protein casein. However, no statistically significant difference in Se removal was observed from the organic amendments investigated compared to the case where no carbon or protein source was supplied (Flury et al. (1997).

Fellowes et al. (2013) investigated the effect of microbial activity on Se cycling in the environment for a Se-enriched soil in Ireland. In microcosm experiments under anoxic conditions, supplementation of soil with selenate resulted in rapid reduction to Se(0). In contrast, addition of sodium acetate as electron donor had no effect on the selenate removal. Rapid denitrification was also observed upon addition of nitrate to the microcosm, hindering selenate reduction and reoxidizing reduced forms of Se (Fellowes et al., 2013). Thus, the study discourages inclusion of selenate into nitrogenous fertilizers as this could lead to mobilisation and release of Se in surface run-off. The effect of organic amendments such as poultry manure, sugarcane press mud and farmyard manure on Se uptake and volatilization on seleniferous soils in north-western India has been investigated by Dhillon et al. (2010). Utilization of press mud and poultry manure effectively enhanced volatilization and inhibited transfer of Se from seleniferous soil to the plants. Thus, the deleterious effects of Se were alleviated and the nutritional quality of the grains restored (Dhillon et al., 2010; Sharma et al., 2011).

In the present work, the pH drop in the microcosms amended with glucose and maize husk (**Figure 4.3**) could be attributed to the significant production of volatile fatty acids (mainly acetate) as observed in **Figures 4.2c and 4.2d**. However, the drop in pH did not affect the Se reduction efficiency by the soil microbial community in the microcosms. Bioreduction of Se oxyanions under acidic conditions, typical for acid mine drainage, is difficult due to the lower efficiency of the bacterial metabolism at low pH (Lenz et al., 2008). Based on these findings, it is recommended to isolate microorganisms capable of Se reduction at low pH. The microorganisms may be suitable for bioaugmentation applications in large-scale reactors treating Se contaminated acid mine drainage.

4.4.2. Biological treatment of soil leachate in a UASB reactor

Efficient Se removal (> 90%) was obtained by the UASB reactor resulting in a Se concentration lower than 5 μg L^{-1} in the treated effluent (**Figure 4.4d**), which is in accordance with the Se discharge limit set by the USEPA (Tan et al., 2016). However, it may be noted that the Se-leachate obtained from soil washing and fed to the UASB reactor contained a relatively low Se concentration, around 55 μg L^{-1} (**Figure 4.4d**). Lenz et al. (2008) obtained 90% removal of 780 μg Se L^{-1} provided as selenate in methanogenic and sulfate-reducing UASB reactors using granular sludge as inoculum from the same full-scale reactor as in the present study. Moreover, the presence of metals, organic contaminants or other impurities in the influent may also have influenced the Se removal efficiency from the soil leachate. Luo et al. (2008) achieved a selenate decrease from 200 μg L^{-1} to less than 50 μg L^{-1} in the presence of co-contaminants such as arsenate and sulfate using an ethanol-fed sulfate-reducing packed-bed bioreactor. Bioreduction of 79 mg L^{-1} selenate in the presence of cadmium, lead and zinc as co-contaminants not only resulted in high Se reduction and removal efficiencies (92%) from artificial wastewater, but also in biogenic formation of quantum dots (Mal et al., 2016, 2017a).

In addition to the anaerobic granular sludge activity, the indigenous soil microorganisms might have enhanced the Se removal efficiency of the UASB reactor. Soil microorganisms obtained from seleniferous regions have been reported to harbour an enhanced Se reduction capacity. Anaerobic bacteria such as *Azospirullum brasilense* (Tugarova et al., 2014) and *Paenibacillus selenitireducens* (Yao et al., 2014) and aerobic bacteria such as *Bacillus* sp. (Fujita et al., 1997) and *Rhizobium selenireducens* (Hunter, 2014) isolated from Se-contaminated soils showed capability to convert soluble Se to elemental Se(0).

In phase IV of reactor operation, the electron donor supplementation was gradually reduced while in phase V, the lactate supply was completely stopped. However, the organic content in the Se-leachate from seleniferous soil washing provided the electrons required for Se bioreduction by the granular sludge (**Figure 4.4**). The organic content in the soil leachate was analysed by 3DEEM (**Figure 4.5**). The intensity related to fulvic and humic substances in the influent was lower than that of the effluent. This can be attributed to the release of EPS components from the granular sludge to the treated leachate. Sulfate reduction in an inversed fluidized bed bioreactor inoculated with granular sludge from the same full scale reactor was still observed although supplementation of lactate as electron donor was stopped (Cassidy et al., 2017). The study suggested formation of polyhydroxybutyrate (PHB) storage products by the microbial community in the granular sludge during reactor operation, which was utilized as electron donor for microbial processes during organic-deprived conditions.

EDS and P-XRD analysis showed the deposition of Se(0) onto the surface of the UASB granules. This observation was confirmed by the presence of a peak at 34.34° (**Figure 4.7**), characteristic of hexagonal Se, on the granules at days 45 and 78, and its absence on the granules at day 0. Several studies of Se bioreduction processes (Jain et al., 2015; Mal et al., 2017b) using the same granular sludge reported that most of the reduced Se deposited on the EPS of the granular sludge. The recovery of the Se from the EPS can be achieved by centrifugation or sonication techniques (Basuvaraj et al., 2015; Mal et al., 2017b).

4.4.3. Soil leachate treatment

Seleniferous soils from the regions of Punjab (India) have induced Se toxicity symptoms in the local population. The source of Se in the soil is from the mountain ranges in the north (Siwalik hills), and is constantly deposited in the agricultural soil of Punjab via rivulets (Dhillon et al., 2008). Since the area is primarily used for agricultural purposes, it is necessary to remediate the land in order to decrease the contamination. Moreover, Se recovery in order to meet the increasing demand for Se in the electronic and healthcare industries (Lenz and Lens, 2009) could provide useful assistance to balance the cost of soil remediation.

Soil washing provides a permanent treatment alternative to achieve contaminant removal from soil. It is a pre-concentration step which allows contaminant extraction using a washing solution, which is then further treated before discharge (Dermont et al., 2008). Soil washing for remediation of heavy metal and hydrocarbon contaminated soil leachate has been widely studied (Ferraro et al., 2015; Huguenot et al., 2015; Mousset et al., 2016; Trellu et al., 2016). Various physico-chemical treatments such as adsorption and photocatalysis for soil washing effluent containing Cu, Zn, Fe and Mn (Satyro et al., 2016), alkali treatment for Cd-contaminated soil leachate (Makino et al., 2016), electrochemical treatment for Pb, Zn, Cd and As contaminated soil leachate (Pociecha and Lestan, 2012), coagulation/precipitation of As in soil washing effluent (Jang et al., 2005) and immobilisation of Pb in the leachate by application of calcite and allophanic soil (Isoyama and Wada, 2007) have been applied for treatment of the washing solution after remediation of heavy metals contaminated soils.

Other studies applied biological techniques for soil leachate treatment. Emenike et al. (2017) studied the removal of Pb, Al and Cu from soil leachate using bioaugmentation with three bacterial species (*Lysinibacillus* sp., *Bacillus* sp. and *Rhodococcus* sp.). A further case study showed that the native microbial community in a heavy metal (Cu and Cd) contaminated soil formed a periphytic biofilm which optimally removed the heavy metals from the acidic soil leachate (Wu et al., 2017). Thus, biological treatment provides a cost-effective, environmental-friendly and sustainable alternative to the expensive physico-chemical techniques for soil leachate remediation.

4.5. Conclusions

The indigenous soil microbiota and the organic carbon content were found to be sufficient to achieve Se reduction in the investigated soil under optimal environmental conditions. *In situ* soil remediation did not completely remove Se from the soil and it only converted mobile Se forms to immobilized Se. In contrast, the soil washing technique completely removed Se from the soil environment and avoids future possibilities of re-oxidation and mobilization. This study demonstrated that soil washing using environmental friendly washing agents (i.e. water) and biological treatment of Se-rich soil leachate are promising approaches for Se removal and recovery from Se-rich soil. This *ex situ* approach allowed to remove Se from the seleniferous soil and retain Se in the form of biogenic Se(0) in the granular sludge which may be recovered for re-use. Along with the granular sludge, the indigenous soil microorganisms as well as the high organic carbon content of the soil offer a low-cost and efficient bioreduction solution for seleniferous soil treatment.

References

APHA/AWWA/WEF (2012) Standard methods for the examination of water and wastewater, 22nd ed, Standard Methods. American Public Health Association, American Water Works Association, Water Environment Federation. Washington

Ashworth DJ, Shaw G (2006) Soil migration, plant uptake and volatilisation of radio-selenium from a contaminated water table. Sci Total Environ 370(2-3):506–514

Basuvaraj M, Fein J, Liss SN (2015) Protein and polysaccharide content of tightly and loosely bound extracellular polymeric substances and the development of a granular activated sludge floc. Water Res 82:104–117

Cassidy J, Frunzo L, Lubberding HJ, Villa-Gomez DK, Esposito G, Keesman KJ, Lens PNL (2017) Role of microbial accumulation in biological sulphate reduction using lactate as electron donor in an inversed fluidized bed bioreactor: Operation and dynamic mathematical modelling. Int Biodeterior Biodegrad 121:1–10

Cennamo P, Montuori N, Trojsi G, Fatigati G, Moretti A (2016) Biofilms in churches built in grottoes. Sci Total Environ 543:727-738

Chen W, Westerhoff P, Leenheer JA, Booksh K (2003) Fluorescence excitation - Emission matrix regional integration to quantify spectra for dissolved organic matter. Environ Sci Technol 37(24):5701–5710

De La Riva DG, Vindiola BG, Castañeda TN, Parker DR, Trumble JT (2014) Impact of selenium on mortality, bioaccumulation and feeding deterrence in the invasive Argentine ant, *Linepithema humile* (Hymenoptera: Formicidae). Sci Total Environ 481:446–452

Dermont G, Bergeron M, Mercier G, Richer-Laflèche M (2008) Soil washing for metal removal: A review of physical/chemical technologies and field applications. J Hazard Mater 152(1):1–31

Dhillon KS, Dhillon SK, Dogra R (2010) Selenium accumulation by forage and grain crops and volatilization from seleniferous soils amended with different organic materials. Chemosphere 78(5):548–556

Dhillon SK, Dhillon KS, Kohli A, Khera KL (2008) Evaluation of leaching and runoff losses of selenium from seleniferous soils through simulated rainfall. J Plant Nutr Soil Sci 171(2):187–192

El-Ramady H, Abdalla N, Alshaal T, Domokos-Szabolcsy É, Elhawat N, Prokisch J, Sztrik A, Fári M, El-Marsafawy S, Shams MS (2015) Selenium in soils under climate change, implication for human health. Environ Chem Lett 13(1):1-19

Emenike CU, Liew W, Fahmi MG, Jalil KN, Pariathamby A, Hamid FS (2017) Optimal removal of heavy metals from leachate contaminated soil using bioaugmentation process. Clean - Soil, Air, Water 45(2):1500802

Fellowes JW, Pattrick RAD, Boothman C, Al Lawati WMM, van Dongen BE, Charnock JM, Lloyd JR, Pearce CI (2013) Microbial selenium transformations in seleniferous soils. Eur J Soil Sci 64(5):629–638

Ferraro A, van Hullebusch ED, Huguenot D, Fabbricino M, Esposito G (2015) Application of an electrochemical treatment for EDDS soil washing solution regeneration and reuse in a multi-step soil washing process: Case of Cu contaminated soil. J Environ Manage 163:62–69

Flury M, Frankenberger Jr WT, Jury WA (1997) Long-term depletion of selenium from Kesterson dewatered sediments. Sci Total Environ 198:259–270

Frankenberger Jr WT, Amrhein C, Fan TWM, Flaschi D, Glater J, Kartinen Jr E, Kovac K, Lee E, Ohlendorf HM, Owens L, Terry N, Toto A (2004) Advanced treatment technologies in the remediation of seleniferous drainage waters and sediments. Irrig Drain Syst 18(1):19–42

Fujita M, Ike M, Nishimoto S, Takahashi K, Kashiwa M (1997) Isolation and characterization of a novel selenate-reducing bacterium, *Bacillus* sp. SF-1. J Ferment Bioeng 83(6):517–522

Gadd GM (2000) Bioremedial potential of microbial mechanisms of metal mobilization and immobilization. Curr Opin Biotechnol 11(3):271–279

Hagarova I, Zemberyova M, Bajcan D (2003) Sequential and single step extraction procedures for fractionation of selenium in soil samples. Chem Pap 59(2):93–98

Huguenot D, Mousset E, van Hullebusch ED, Oturan MA (2015) Combination of surfactant enhanced soil washing and electro-Fenton process for the treatment of soils contaminated by petroleum hydrocarbons. J Environ Manage 153:40-47

Hunter WJ (2014) A *Rhizobium selenitireducens* protein showing selenite reductase activity. Curr Microbiol 68(3):311–316

Isoyama M, Wada SI (2007) Remediation of Pb-contaminated soils by washing with hydrochloric acid and subsequent immobilization with calcite and allophanic soil. J Hazard Mater 143(3):636–642

Jain R, Jordan N, Weiss S, Foerstendorf H, Heim K, Kacker R, Hubner R, Kramer H, Van Hullebusch ED, Farges F, Lens PNL (2015) Extracellular polymeric substances govern the surface charge of biogenic elemental selenium nanoparticles. Environ Sci Technol 49(3):1713–1720

Jang M, Jung SH, Sang IC, Jae KP (2005) Remediation of arsenic-contaminated soils and washing effluents. Chemosphere 60(3):344–354

Kabata-pendias A, Pendias H (2001) Trace elements in soils and plants trace elements in soils and plants, 3rd ed. CRC press LLC, Boca Raton, Florida

Lenz M, Lens PNL (2009) The essential toxin: the changing perception of selenium in environmental sciences. Sci Total Environ 407(12):3620-3633

Lenz M, van Hullebusch ED, Hommes G, Corvini PFX, Lens PNL (2008) Selenate removal in methanogenic and sulfate-reducing upflow anaerobic sludge bed reactors. Water Res 42(8-9):2184–2194

Lindblom SD, Fakra SC, Landon J, Schulz P, Tracy B, Pilon-Smits EAH (2014) Inoculation of selenium hyperaccumulator *Stanleya pinnata* and related non-accumulator *Stanleya elata* with hyperaccumulator rhizosphere fungi - investigation of effects on Se accumulation and speciation. Physiol Plant 150(1):107–118

Luo Q, Tsukamoto TK, Zamzow KL, Miller GC (2008) Arsenic, selenium, and sulfate removal using an ethanol-enhanced sulfate-reducing bioreactor. Mine Water Environ 27(2):100–108

Luongo V, Ghimire A, Frunzo L, Fabbricino M, d'Antonio G, Pirozzi F, Esposito G (2017) Photofermentative production of hydrogen and poly-β-hydroxybutyrate from dark fermentation products. Bioresour Technol 228:171-175

Makino T, Maejima Y, Akahane I, Kamiya T, Takano H, Fujitomi S, Ibaraki T, Kunhikrishnan A, Bolan N (2016) A practical soil washing method for use in a Cd-contaminated paddy field, with simple on-site wastewater treatment. Geoderma 270:3–9

Mal J, Nancharaiah YV, Bera S, Maheshwari N, van Hullebusch ED, Lens PNL (2017a) Biosynthesis of CdSe nanoparticles by anaerobic granular sludge. Environ Sci Nano 4:824–833

Mal J, Nancharaiah YV, van Hullebusch ED, Lens PNL (2016) Effect of heavy metal co-contaminants on selenite bioreduction by anaerobic granular sludge. Bioresour Technol 206:1–8

Mal J, Nancharaiah YV, Maheshwari N, van Hullebusch ED, Lens PNL (2017b) Continuous removal and recovery of tellurium in an upflow anaerobic granular sludge bed reactor. J Hazard Mater 327:79–88

Misra S, Boylan M, Selvam A, Spallholz J, Björnstedt M (2015) Redox-active selenium compounds - from toxicity and cell death to cancer treatment. Nutrients 7(5):3536–3556

Moreno RG, Burdock R, Cruz M, Álvarez D, Crawford JW (2013) Managing the selenium content in soils in semiarid environments through the recycling of organic matter. Appl Environ Soil Sci 2013:283468

Mousset E, Huguenot D, van Hullebusch ED, Oturan N, Guibaud G, Esposito G, Oturan MA (2016) Impact of electrochemical treatment of soil washing solution on PAHdegradation efficiency and soil respirometry. Environ Pollut 211:354-362

Oldfield JE (2002) Se World atlas. Selenium-tellurium development association, Grimbergen, Belgium

Pociecha M, Lestan D (2012) Recycling of EDTA solution after soil washing of Pb, Zn, Cd and As contaminated soil. Chemosphere 86(8):843–846

Pontoni L, van Hullebusch ED, Pechaud Y, Fabbricino M, Esposito G, Pirozzi F (2016) Colloidal mobilization and fate of trace heavy metals in semi-saturated artificial soil (OECD) irrigated with treated wastewater. Sustainability 8(12):1257

Roest K, Heilig HG, Smidt H, de Vos WM, Stams AJ, Akkermans AD (2005) Community analysis of a full-scale anaerobic bioreactro treating paper mill wastewater. Syst Appl Microbiol 28(2):175-185

Satyro S, Race M, Di Natale F, Erto A, Guida M, Marotta R (2016) Simultaneous removal of heavy metals from field-polluted soils and treatment of soil washing effluents through combined adsorption and artificial sunlight-driven photocatalytic processes. Chem Eng J 283(1):1484–1493

Sharma S, Bansal A, Dogra R, Dhillon SK, Dhillon KS (2011) Effect of organic amendments on uptake of selenium and biochemical grain composition of wheat and rape grown on seleniferous soils in northwestern India. J Plant Nutr Soil Sci. 174(2):269–275

Stams AJM, Grolle KCF, Frijters CTMJ, Van Lier JB (1992) Enrichment of thermophilic propionate-oxidizing bacteria in syntrophy with *Methanobacterium thermoautotrophicum* or *Methanobacterium thermoformicicum*. Appl Environ Microbiol 58(1):346–352

Tan LC, Nancharaiah YV, van Hullebusch ED, Lens PNL (2016) Selenium: Environmental significance, pollution, and biological treatment technologies. Biotechnol Adv 34(5):886–907

Trellu C, Ganzenko O, Papirio S, Pechaud Y, Oturan N, Huguenot D, van Hullebusch ED, Esposito G, Oturan MA (2016) Combination of anodic oxidation and biological treatment for the removal of phenanthrene and Tween 80 from soil washing solution. Chem Eng J 306:588-596

Tugarova AV, Vetchinkina EP, Loshchinina EA, Burov AM, Nikitina VE, Kamnev AA (2014) Reduction of selenite by *Azospirillum brasilense* with the formation of selenium nanoparticles. Microb Ecol 68(3):495–503

Wadgaonkar SL, Ferraro A, Race M, Nancharaiah YV, Dhillon KS, Fabbricino M, Esposito G, Lens PNL (2017) Optimisation of soil washing to reduce selenium level of seleniferous soil from Punjab, Northwestern India. Chemosphere (submitted)

Wang Q, Zhang J, Zhao B, Xin X, Deng X, Zhang H (2016) Influence of long-term fertilization on selenium accumulation in soil and uptake by crops. Pedosphere 26(1):120–129

Williams KH, Wilkins MJ, N'Guessan AL, Arey B, Dodova E, Dohnalkova A, Holmes D, Lovley DR, Long PE (2013) Field evidence of selenium bioreduction in a uranium-contaminated aquifer. Environ Microbiol Rep 5(3):444–452

World Health Organization (2011) Selenium in drinking-water. Geneva

Wu Y, Yang J, Tang J, Kerr P, Wong PK (2017) The remediation of extremely acidic and moderate pH soil leachates containing Cu (II) and Cd (II) by native periphytic biofilm. J Clean Prod 162:846–855

Yao R, Wang R, Wang D, Su J, Zheng S, Wang G (2014) *Paenibacillus selenitireducens* sp. nov., a selenite-reducing bacterium isolated from a selenium mineral soil. Int J Syst Evol Microbiol 64(3):805–811

Yasin M, El Mehdawi AF, Jahn CE, Anwar A, Turner MFS, Faisal M, Pilon-Smits EAH (2014) Seleniferous soils as a source for production of selenium-enriched foods and potential of bacteria to enhance plant selenium uptake. Plant Soil 386(1-2):385–394

CHAPTER 5

Phytoremediation of seleniferous soil leachate using the
aquatic plants *Lemna minor* and *Egeria densa*

This chapter has been modified and published as:

Ohlbaum M, Wadgaonkar SL*, van Bruggen H, Nancharaiah YV, Esposito G, Lens PNL.
(2018) Phytoremediation of seleniferous soil leachate using the aquatic plants *Lemna minor*
and *Egeria densa*. Ecol. Eng. 120: 321-328. DOI: 10.1016/j.ecoleng.2018.06.013
(*corresponding author)

Abstract

Phytoremediation of selenium (Se)-containing Hoagland solution and seleniferous soil leachate was evaluated using with two aquatic plants, *Lemna minor* and *Egeria densa*. The soil leachate was prepared by washing the soil in a liquid to solid (L:S) ratio of 10:1 (v/w) with H_2O, $KMnO_4$ and $K_2S_2O_8$ at 250 rpm for 6 h. The composition of each seleniferous soil leachate was determined. Based on this, an artificial seleniferous soil leachate was modelled for understanding the phytoremediation process. *L. minor* showed the highest Se removal efficiency (97%) in Hoagland solution, followed by 85% Se removal in the artificial soil leachate. *L. minor* showed removal efficiencies of 80% and 99% for Se and Mn, respectively, due to simultaneous removal of both the metal(loid)s. Addition of $K_2S_2O_8$, however, decreased the uptake of Se by the plants investigated by 40% and the medium pH from 7 to 3. Se removal efficiency of aquatic plants decreased by 30% when sulfate was included. *L. minor* showed a Se removal efficiency of 76% with Se uptake of 29 μg g^{-1} dry weight from seleniferous soil leachate which contained 74 μg L^{-1} Se. This study demonstrated that aquatic plants such as *Lemna minor* and *Egeria densa* were able to remove Se from seleniferous soil leachate and phytoremediation efficiency depends on the composition of the extractant used for soil washing.

Keywords: Soil washing, phytoremediation, selenium, aquatic plants, seleniferous soil, soil leachate

5.1. Introduction

Selenium (Se) is an essential trace element as it plays an indispensable role in the functioning of oxidoreductase enzymes in humans and animals (Hatfield et al. 2014).). Although it is not considered essential for plants, it has reportedly shown positive effects on the growth and resistance towards pathogens and herbivores (Yasin et al., 2014; Handa et al., 2015). In adult human diet, the recommended daily dose for Se is 0.04 - 0.4 mg (FAO/WHO, 2001). In Se deficient regions such as Finland and Mediterranean countries, Se is commonly added as inorganic fertilizer to crops to achieve optimal levels of Se in the diet (Vita et al., 2016).

Excess Se in human diet has, on the contrary, been reported to cause Se toxicity which leads to skin diseases and gastrointestinal and neurological symptoms (Sunde et al., 2012). Therefore, Se toxicity is one of the major concerns of regions in north India, China and Australia, where selenium levels in soil are high. According to Dhillon and Dhillon (2003), soils containing more than 0.5 mg Se kg^{-1} soil are considered as seleniferous. Such soil used as agricultural land may directly influence the dietary intake of the local population (Hira et al., 2004).). Sources of selenium in soil may vary, ranging from lithogenic i.e. formation of soil from parent rock (Haug et al. 2007) or anthropogenic activities such as mining and agriculture (Winkel et al., 2011).

Treatment of seleniferous soils by the soil washing technique has several advantages over conventional soil bioremediation techniques such as bioamendment or phytoremediation (Wuana and Okiemen, 2011). Soil washing is a fast and cost effective procedure that decreases the volume of soil that needs to be treated (Dermont et al., 2008). Aquatic plants are usually invasive species that require low maintenance to live and have a high biomass growth (Crites et al. 2014; Wu et al. 2015). These plants are known to be tolerant to mining and industrial wastewaters and are able remove organic and inorganic contaminants. Aquatic plants can also take up and accumulate heavy metals in their tissues (Marchand et al., 2010). Recent studies have shown that aquatic plants are also able to remove and take up Se from solutions as either SeO_3^{2-} or SeO_4^{2-} (Mechora et al, 2011). Thus, application of phytoremediation using aquatic plants for removing metals from seleniferous soil leachate would lead to development of a sustainable process for remediation of Se-contaminated soils.

In a previous study, a procedure involving oxidising agents such as $KMnO_4$ and $K_2S_2O_8$, used for soil washing showed the highest efficiency for removing Se from seleniferous soil (**Chapter 3**). However, the chemical nature of the leachate might adversely affect the soil and may increase the treatment costs of the remediation. This study for the first time, combines soil washing for removing Se from seleniferous soil and phytoremediation for removing Se from the soil leachate. The aim of this research was to evaluate the efficiency of *Lemna minor* and *Egeria densa* for the removal and uptake of Se from seleniferous soil leachate. Different parameters such as Se concentration, composition of the artificial and real soil leachate and effect of oxidising agents on Se removal and Se uptake by plants have been evaluated.

5.2. Materials and methods

5.2.1. Sample collection and storage

Soil was collected from the northwest region of India, in the state of Punjab. The geographical co-ordinates of the sampling location was 31° 07' 45.5"N; 76° 12' 43.1"E. The soil was collected from a sampling depth of 12 cm of the agricultural region. The soil was air-dried, sieved and stored at room temperature for further analysis.

Aquatic plants, *L. minor* and *E. densa* were collected from a private aquarium and the Aquariumhuis (Romberg, the Netherlands), respectively. The plants were grown in 5 L buckets with diluted Hoagland solution in order to adapt to the nutrient media and the laboratory greenhouse conditions for two weeks before exposure to selenium containing effluent. The composition of the Hoagland solution was adapted from Megateli et al. (2009) as follows (in mg L^{-1}): $Ca(NO_3)_2 \cdot 4H_2O$ (118), KNO_3 (5), $MgSO_4 \cdot 7H_2O$ (5), KH_2PO_4 (0.68), $FeSO_4 \cdot 7H_2O$ (0.3), K_2SO_4 (0.35), H_3BO_3 (0.3), $MnSO_4 \cdot 7H_2O$ (0.15), $ZnSO_4$ (0.022), $CuSO_4$ (0.008), $NiSO_4 \cdot 7H_2O$ (0.005). All analytical-grade chemicals used for soil washing were purchased from Merck. The plants were then harvested to assess the effect of Se exposure under varying conditions.

5.2.2. Soil washing

The soil washing procedure was performed according to the procedures mentioned in chapter 3 of this thesis. The 2 g soil was washed with 20 mL of each 0.1% oxidising agent ($KMnO_4$ and $K_2S_2O_8$) at ambient temperature at 150 rpm in a rotary shaker (Model: INNOVA 2100, New Brunswick Scientific, New Jersey, USA) for 6 h. 2 g soil was washed with 20 mL milliQ (MQ) water as control. After washing, the solutions were centrifuged at 3000 rpm at 20 °C and the supernatant was filtered through 0.45 μm cellulose acetate filters. The composition of the filtrate was analysed to simulate artificial soil leachate for further studies.

5.2.3. Experimental set-up

Evaluation of Se removal by the aquatic plants, *L. minor* and *E. densa*, were exposed to varying Se concentrations, time of Se exposure, composition of washing solution containing varying concentrations of $MnSO_4$, $K_2S_2O_8$, and SO_4^{2-} and real effluent obtained from washing seleniferous soil.

For each test, 1 g (wet weight) of each plant was exposed to 100 mL of solution in 135 mL transparent plastic cups. The temperature in the greenhouse varied between 25-30 °C and light was provided with a minimum light intensity of 100 µmol $m^{-2}s^{-1}$ photons. All the experiments were performed in duplicates. After one week of exposure and daily visual inspection, the plants were harvested and dried at 70 °C overnight for further analysis. To control losses through evaporation, the volume of the solution of each bucket was measured after the experiment. The effluent solutions were sampled at day 0 and day 7 for Se analysis. The Se removal efficiency (%) was calculated by the following equation:

$$\text{Removal efficiency (\%)} = \frac{\left[\text{Se initial}\left(\frac{\mu g}{L}\right) - \text{Se final}\left(\frac{\mu g}{L}\right)\right]*100}{\text{Se initial}\left(\frac{\mu g}{L}\right)} \qquad \text{(Eq. 5.1)}$$

A preliminary experiment was performed to evaluate the effect of high Se concentrations on *L. minor* and *E. densa*. The plants were exposed to 50, 100 and 500 mg L^{-1} sodium selenate in Hoagland solution and the Se removal efficiency was evaluated. Moreover, the plants were exposed to the artificial soil leachate simulated from the soil leachate analysis.

In order to acclimatize the plants to the soil washing medium, the plants were grown in the artificial soil washing medium (without Se) prior to exposure to the simulated soil leachate with oxidising agents and Se. The simulated soil leachate containing different concentrations (in mg L^{-1}) of the residual components of oxidising agents such as Mn (0.5, 1, 2), $K_2S_2O_8$ (1, 5, 100, 500) and SO_4^{2-} (50, 100, 500) each containing 100 µg L^{-1} selenium were used as effluent and exposed to the individual plants. A validation experiment was done using 1 g of *L. minor* exposed during 7 days to 100 mL of a real soil leachate of the soil washed with only water.

5.2.4. Analysis

Total Se was measured by atomic absorption spectrophotometer coupled with graphite furnace (AAS-GF, Thermos elemental solar MQZ, GF95, England) as described by Mal et al. (2016). Heavy metals and elements were measured by Inductively Coupled Plasma Mass Spectrometry

(ICP-MS, XSeriesII, Thermo Scientific, Germany). Anions were measured using a Dionex ICS-1000 Ion Chromatograph (Thermo Fisher Scientific, Breda, Netherlands) as described by Dessi et al. (2016).

Wet weight (g) and growth (%) of plants was measured at the end of the experiment. Negative growth (%) values indicate loss of plants material due to deleterious growth conditions. Growth (%) was calculated using the following equation:

$$\text{Growth (\%)} = \frac{(\text{Final weight} - \text{Initial weight}) * 100}{\text{Initial weight}} \qquad \text{(Eq. 5.2)}$$

Chlorophyll a was measured using the spectrophotometer method (Wintermand and de Mots 1965). Chlorophyll a was extracted with 5 mL of ethanol (96%) in centrifuges tubes. After 12 hours, the samples were centrifuged at 3000 rpm for 10 min until a clear supernatant was obtained. The absorbance was measured at 649, 655 and 750 nm using a UV-Vis spectrophotometer (PerkinElmer Lambda 365 UV/Vis, Groningen, the Netherlands) to determine the chlorophyll-a, according to the following equation:

$$\text{Chlorophyll a} = \frac{[13.7 \times (A665 - A750) - 5.76 \times (A649 - A750)] \times V \times D}{W \times 1000} \qquad \text{(Eq. 5.3)}$$

Where: V= extraction volume (mL), D= sample dilution factor, and W= wet weight of sample (g)

Plants were digested using a microwave accelerated reaction system (CEM Mars 5, Abcoude, Netherlands) with 10 mL of HNO_3 at 165 °C for 10 min and 175 °C for 2 min. The solution was suitably diluted, filtered and analysed for total Se. One way analysis of variance ANOVA was applied to evaluate Se uptake in plants and plant parameters using the open software R studio (at 95% confidence levels).

5.3. Results

5.3.1. Soil washing analysis

The highest concentration of Se (130 ± 3.5 µg L^{-1}) was observed in the effluent of the soil washed with $KMnO_4$ (**Table 5.1**). All the three soil leachates contained low levels of heavy metals such as As, Cd and Pb. Around 20-fold increase in Cr concentration was observed when the soil was washed with $KMnO_4$ compared to that of control. The Mn concentration in the $KMnO_4$ washed soil leachate was around 2.5 times higher than that of the control and was almost similar to that of the $K_2S_2O_8$ effluent. While washing the soil with $K_2S_2O_8$, the SO_4^{2-}

concentration significantly increased from 5 mg L^{-1} in water washed soil leachate to 150 mg L^{-1} in $K_2S_2O_8$ washed soil leachate. The levels of K, Ca and Cl$^-$ were also higher in the $K_2S_2O_8$ washed soil leachate than those of H_2O and $KMnO_4$ washed soil leachates.

Table 5.1. Macro and trace element composition of the soil leachates using MQ water, $KMnO_4$ and $K_2S_2O_8$ as washing solutions

Macro elements (mg L^{-1})	MQ water	$KMnO_4$	$K_2S_2O_8$
Na	6.09 ± 0.07	6.42 ± 0.34	7.10 ± 0.62
Mg	6.64 ± 0.02	7.29 ± 0.21	18.02 ± 0.49
Al	0.69 ± 0.08	0.13 ± 0.04	0.057 ± 0.02
K	6.18 ± 1.65	88.8 ± 1.63	130.81 ± 7.04
Ca	11.76 ± 0.06	8.54 ± 0.38	33.49 ± 0.67
Fe	0.88 ± 0.22	0.22 ± 0.10	0.065 ± 0.04
Cl$^-$	13.11 ± 1.02	4.06 ± 0.66	28 ± 0.05
NO$_3^-$	1.73 ± 0.75	1.69 ± 0.68	1.1 ± 0.20
SO$_4^{2-}$	3.77 ± 0.28	5.58 ± 0.31	151.41 ± 8.26
PO$_4^{3-}$	0.45 ± 0.03	0.38 ± 0.02	0
Trace elements (µg L^{-1})			
Se	43.0 ± 1.4	130 ± 3.5	90.0 ± 0.8
Mn	20.55 ± 2.47	46.80 ± 19.66	39.95 ± 37.55
Cr	1.60 ± 0.14	39.60 ± 0.14	0.95 ± 0.21
Co	0.35 ± 0.07	0.35 ± 0.07	0.45 ± 0.35
Ni	2.25 ± 0.07	1.75 ± 0.21	1.55 ± 0.07
Cu	22.20 ± 0.28	28.80 ± 3.68	18.50 ± 1.98
Zn	12.00 ± 0.28	9.60 ± 3.11	16.25 ± 6.43
Cd	<0.1	0.13 ± 0.04	0.10 ± 0.14
Pb	0.70 ± 0.14	0.3	0.10 ± 0.14
As	13.7 ± 2.4	4.0 ± 0.3	3.5 ± 1.2

The artificial soil leachate was prepared (**Table 5.2**) to mimic the composition of the washed-soil solution of the soil washed with MQ water, $KMnO_4$ or $K_2S_2O_8$. Heavy metals were excluded from the artificial soil leachate in order to study exclusively the effect of the oxidising agents on the selenium removal efficiency by the aquatic plants. The pH of the artificial soil leachate was maintained at 7 by the addition of carbonate as Na_2CO_3.

Table 5.2. Chemical composition of artificial soil leachate

Salt composition	Concentration (mg L^{-1})
KNO_3	6.50
Na_2CO_3	3.00
$MgSO_4$	4.00
$MgCl_2 \cdot 6 H_2O$	45.0
KH_2PO_4	0.40
$CaCl_2$	30.0
$FeSO_4 \cdot 7 H_2O$	1.0
$MnSO_4 \cdot 7 H_2O$	0.10
$NiSO_4 \cdot 7 H_2O$	0.01
$CuSO_4 \cdot 5 H_2O$	0.09
$ZnSO_4 \cdot 7 H_2O$	0.06

5.3.2. Selenium removal at different concentrations

L. minor showed the highest efficiency for Se removal (~97.8%) when exposed to 50 µg L^{-1} Se in Hoagland solution, followed by *E. densa* which showed ~87.6% removal at the same Se concentration. In both plants, the removal efficiency decreased with an increase in Se concentration in the solution (**Figure 5.1a**). However, an increase in Se uptake was observed in both the plants with increasing exposure to Se concentrations in the Hoagland solution. The highest Se uptake was 69.09 (±1.86) µg g^{-1} dry weight and 43.40 (±1.47) µg g^{-1} dry weight by *L. minor* and *E. densa* respectively when exposed to 500 g L^{-1} Se in Hoagland solution. After 7 days of Se treatment, the plant chlorophyll a content had diminished with increasing Se concentration (**Figure 5.1b**).

The chlorophyll a content (**Figure 5.2a**) decreased from 0.36 (±0.04) to 0.13 (±0.20) mg g^{-1} fresh weight for *L. minor* and from 0.69 (±0.04) to 0.52 (±0.20) mg g^{-1} fresh weight for *E. densa* when increasing the Se concentration from 50 to 500 µg L^{-1}. Growth analysis of the plants (**Figure 5.2b**) showed that highest increase in growth was observed in *L. minor* by 13.80 (±1.49) % when 50 µg L^{-1} of Se was added, but 500 µg L^{-1} of Se were toxic for the plants, as approximately 5% of the plants died.

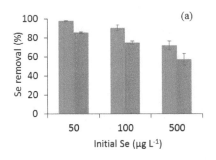 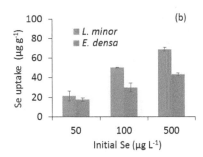

Figure 5.1. Se removal efficiency (a) and Se uptake by plants (b) after the exposure to different Se concentrations in Hoagland solution

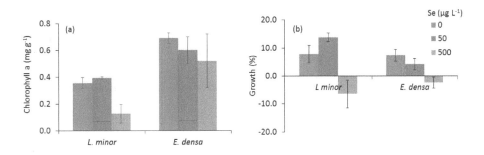

Figure 5.2. Chlorophyll a content (a) and plant growth (b) of *L. minor* and *E. densa* after exposure to different Se concentrations in Hoagland solution

5.3.3. Effect of Mn, K$_2$S$_2$O$_8$ and SO$_4$

Selenium removal

When Mn and Se were present in soil leachate, both plants took up Mn and Se simultaneously with only a slight decrease in the Se removal efficiency with increasing Mn concentration in the solutions (**Figure 5.3a**). The pH of the samples varied between 5.8 and 6.4. *L. minor* showed the highest Se removal efficiency (~86%) for control experiments, where Mn was not added. Addition of 2 mg L^{-1} MnSO$_4$ decreased the Se removal efficiency to 76.8%. A similar trend was observed for *E. densa*, where the selenium removal efficiency decreased from 59.3% to 49.1% with increase in MnSO$_4$ concentration in the medium from 0 to 2 mg L^{-1} (**Figure 5.3a**).

The highest uptake of Se (17.10 ± 1.56 µg g^{-1} dry weight) was observed when *L. minor* was exposed to 1 mg L^{-1} MnSO$_4$ (**Figure 5.3b**). Se uptake by *E. densa* was significantly lower than that of *L. minor*. For *E. densa*, the highest Se uptake (8.01 ±1.62 µg g⁻¹ dry weight) was observed when the plants were exposed to 0.5 mg L⁻¹ MnSO₄ (**Figure 5.3b**).

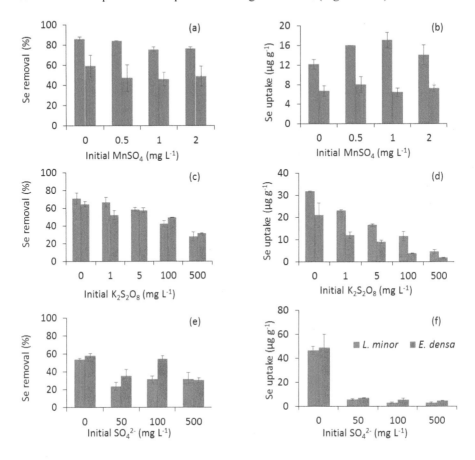

Figure 5.3. Se removal efficiency and uptake of Se by plants after exposure to different concentrations of chemical agents (MnSO₄, K₂S₂O₈ and SO₄²⁻) in artificial soil leachate

The exposure of plants to persulfate (K₂S₂O₈) decreased the Se removal efficiency of both *L. minor* and *E. densa* (**Figure 5.3c**). The addition of K₂S₂O₈ also decreased the pH of the samples, within a range of 3.0-5.8. Increasing persulfate concentration from 0 to 500 mg L⁻¹, decreased the Se removal efficiency from ~71.2% to ~27.9% in *L. minor* and ~65.0% to ~31.4 % in *E. densa*, respectively. Se uptake by both plants considerably decreased with increasing persulfate concentration in the growth medium. The highest Se uptake for *L. minor* and *E. densa* was ~31.7 and ~20.1 µg g⁻¹ dry weight, respectively, in the control solution without addition of persulfate (**Figure 5.3d**). Se uptake was reduced by ~85.8% and ~91.9% for, respectively, *L. minor* and *E. densa* when exposed to 500 mg L^{-1} sulfate in artificial soil leachate (**Figure 5.3d**).

The presence of sulfate also had an inhibitory effect on the Se removal efficiency, as the efficiency decreased from ~57.9% to ~30.3% for *E. densa* and from ~53.6% to ~31.9% in *L. minor* when the plants were exposed from 0 to 500 mg L^{-1} sulfate respectively (**Figure 5.3e**). The effect of sulfate was even more notorious on the uptake of Se by plants. The Se uptake decreased by ~88% for *L. minor* and ~85.8% for *E. densa* upon addition of 50 mg L^{-1} sulfate, while it was further lowered to ~94% and ~90.7% for *L. minor* and *E. densa*, respectively, upon addition of 500 mg L^{-1} sulfate (**Figure 5.3f**). Changes in pH, however, were not as low as with persulfate, and the samples varied from 5.2-6.4.

Plant growth and chlorophyll a content
A slight increase in the growth and chlorophyll a content was observed for *E. densa* in the presence of Mn. For *L. minor*, Mn showed no significant change in the chlorophyll a content, while growth of plants showed a slight decrease (**Figure 5.4a and 5.4b**). However, no statistically significant differences in the growth and chlorophyll a content of both plants were observed. Mn was not detected in the spent effluent solution after 7 days, indicating that plants were able to remove both Mn and Se simultaneously.

Persulfate had a negative effect on plant growth parameters, as ~10% of *L. minor* plants died in the presence of 5 mg L^{-1} $K_2S_2O_8$. *E. densa* showed a very slight increase in plant growth at 5 mg L^{-1} $K_2S_2O_8$, around 10% decrease in plant growth was observed when exposed to 100 mg L^{-1} $K_2S_2O_8$ (**Figure 5.4d**). For the chlorophyll a content, however, the differences were not significant for *L. minor*. In the case of *E. densa*, the differences were significant between initial chlorophyll a content and treatments with $K_2S_2O_8$ (**Figure 5.4c**).

Sulfate addition did not negatively influence the growth or chlorophyll a content in plants. In fact, sulfate addition showed positive effect on the growth of the *E. densa* (**Figure 5.4f**). Up to 12% increase in growth of *E. densa* was observed in the presence of 100 mg L^{-1} sulfate. The increase in plant growth for *L. minor* was, however, statistically insignificant. Also, the chlorophyll a content decreased after the treatment with sulfate (**Figure 5.4e**), however, also these differences were not statistically significant.

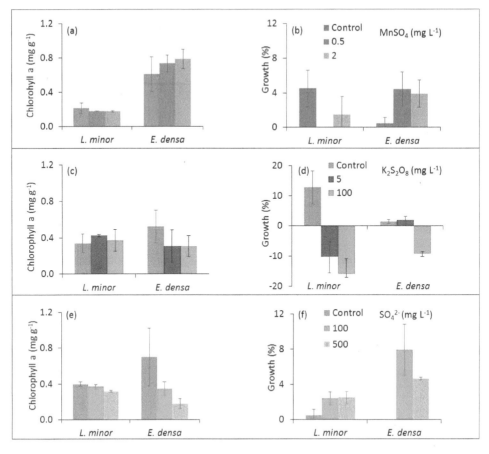

Figure 5.4. Chlorophyll a content and plant growth in *L. minor* and *E. densa* exposed to different concentrations of chemical agents ($MnSO_4$, $K_2S_2O_8$ and SO_4^{2-}) in artificial soil leachate

5.3.4. Phytoremediation of real soil leachate

After washing the soil with water, the Se in the effluent was 74.0 ($\pm$ 1.7) $\mu g \ L^{-1}$. After 7 days of exposure, *L. minor* showed a removal efficiency of 76.0 ($\pm$ 0.3) %, with an uptake of 29.1 ($\pm$ 0.36) $\mu g \ g^{-1}$ dry weight. A Se mass balance was calculated in order to overview the Se distribution over the plants and the leachate. Out of the total initial Se concentration, 30% of the initial Se was taken up by *L. minor*, while 25% of Se persisted in the solution and 50% Se of the total initial Se concentration remained as an unaccounted Se fraction. The unaccounted fraction refers to the difference of the Se in the control (leachate without plants) after 7 days and the total Se in plants and the solution. Since the experiment was carried out in the open system, possible volatilization of Se by aquatic plants might account for the unaccounted Se fraction. Moreover, *L. minor* showed an increase in the growth by 9.6 ($\pm$ 3.4) % in the real soil leachate and 7.9 ($\pm$ 5.8) % in the blank (with artificial soil washing and no Se). The difference in the chlorophyll a content in the initial and final samples in both the real soil leachate and the control was not statistically significant.

5.4. Discussion

5.4.1. Post-treatment of soil and soil leachate

Soil washing has been applied on pilot scale and full-scale for the removal of heavy metals such as Hg, Cr, Cu, Ni, As and Zn and persistent organic compounds such as polychlorinated biphenyl (PCB) and polycyclic aromatic hydrocarbons (PAH) (Lin et al., 2001; Ko et al., 2006). The technique has never been explored for removing Se from seleniferous soils. However, sequential extraction procedures that quantify different Se fractions in soil indicate that Se removal by soil washing is achievable (Supriatin et al., 2015; Schneider et al., 2016). Soil washing is a cost-effective technique and a fast solution for remediation of Se contaminated soils. However, the disadvantage of soil washing is that the final effluent requires a suitable treatment to recover the chemicals and remove heavy metals. The most common techniques applied for post treatment are precipitation, coagulation and ion exchange (Reynier et al., 2015). But, these techniques can be costly and involves the use of chemicals.

Floating and submerged aquatic plants have been studied for the removal of nutrients and heavy metals as they are tolerant to different types of wastewater. For example, *E. densa* and *Hydrilla verticillata* were used to remove As, Zn and Al from a gold mine wastewater effluent (Rezania et al., 2016; Bakar et al. 2013). This study is the first approach to use phytoremediation with aquatic plants for the treatment of the soil leachate contaminated with Se. This would allow to first remove Se from seleniferous soil in a short time, and then treat Se-containing leachate. The plants that are used for removing Se from seleniferous soil leachate can be collected and used as Se fertilizer or for other applications. Aquatic plants have been applied as food for animals or in combination with compost to increase the growth of other plants (van der Siegel et al, 2013; Yao et al, 2017). Other authors have combined phytoremediation and soil washing to improve the soil treatment and metal extraction (Sung et al, 2011; Yan et al. 2017), but so far never to treat directly the effluent.

Analysis of the macro and micro elements in the soil leachate (**Table 5.1**) showed that the application of $KMnO_4$ and $K_2S_2O_8$ can significantly affects the effluent composition. The Se concentration was significantly higher in the leachate with oxidising agents as anticipated.

However, manifold increase in the concentrations of Mg, K, Ca, Cl^-, SO_4^{2-} and Mn in the effluent of the soil washed with $K_2S_2O_8$ and concentrations of K, Mn, Cr in the effluent of the soil washed with $KMnO_4$ was observed. Cr concentration in $KMnO_4$ washed soil leachate was found to be 20 times higher than that of water washed soil leachate. In previous study (**Table 3.2**), less than 0.02 mg kg^{-1} total Cr concentration in the soil was measured. In this study, therefore, it is possible that the Cr increase is due to carryover from the oxidant. In this study, effect of major component of the soil leachates, i.e. the residues of the oxidising agents, on post-treatment of the soil leachate has been studied.

The Mn concentration in the $KMnO_4$ washed soil leachate was significantly lower than in the initial washing solution. This suggests that most of the Mn was precipitated in the soil during the oxidation reaction of $KMnO_4$ with soil elements and organic matter. The high concentration of Mn in soil may adversely affect the agricultural value of the soil. Fertility of this soil for crop production should be assessed in order to determine the feasibility of further use of oxidising agents for Se removal from seleniferous soils.

5.4.2. Phytoremediation of soil leachates

For removing Se, aquatic plants have been mostly studied at the laboratory scale, with nutrient solutions containing only Se, either as SeO_3^{2-} or SeO_4^{2-} (Mechora et al., 2011, Mechora et al., 2015). Carvalho and Martin (2001) observed a very high removal efficiency of SeO_3^{2-} by *Lemna obscura* (95%) and *Hydrilla* (98%). The authors also stated that most of the Se was probably removed as volatile Se compounds after uptake by the plants. In this research, only total Se in the plants was measured after digestion. The highest accumulation of Se in plants was achieved only for the control effluents (**Figure 5.3**), consisting of only 100 µg L^{-1} Se without addition of Mn, sulfate or persulfate. The aquatic plants with high selenium content may be used further as fertilizer in soils low in Se or as animal feed (van der Spiegel et al, 2013; Bañuelos et al. 2015). Biofortification of Se in soils is commonly performed by the addition of inorganic Se to the soil or sprayed into plant leaves, which can reduce the plant yield if the Se doses is not controlled (Ramos et al. 2010), while the application of organic Se would be easier to dosify. However, the amount of Se taken up by aquatic plants is much less than that of hyperaccumulator plants grown on seleniferous soils. Stanleya pinnata, for example, was able to take up to 12900 µg g^{-1} dry weight (Hladun et al., 2011) of selenite.

The high Se removal efficiency of *Hydrilla* (98%) observed by Carvalho et al. (2011) can be compared with the study in Hoagland solution, in which *L. minor* had an average removal of

97% of Se upon exposure to 100 μg L^{-1} Se (**Figure 5.1a**). Hassan and Mostafa (2015) used *Azolla caroliniana* to remove SeO_3^{2-} which was able to take up to 1000 mg kg^{-1} Se when 5 mg L^{-1} Se was added, but at higher Se concentrations, the accumulation decreased and oxygen radicals increased due to Se toxicity towards plants.

Mechora et al. (2011) reported that *Myriophyillum spicatum* accumulated 0.77 μg g^{-1} dry weight upon exposure to 20 μg L^{-1} of Se as SeO_4^{2-} and concentrations of 10 mg L^{-1} resulted in the death of half of the plants, with a highest uptake of 236 (± 7.2) μg g^{-1} dry weight. In this study, up to 40 μg g^{-1} dry weight was taken up by *L. minor* when 100 μg L^{-1} of SeO_4^{2-} was added (**Figure 5.1b**). Mechora et al. (2015) also studied the effect of different SeO_3^{2-} concentrations in *L. minor* and observed that 1 mg L^{-1} can stimulate growth, but at higher concentrations, the plant exhibits chlorosis until 10 mg L^{-1} killing 50% of plants. In this study, the exposure of plants to 0.5 mg L^{-1} of SeO_4^{2-} in Hoagland solutions had a negative effect on the growth of plants and a decrease in the Se removal efficiency. The toxicity could be accounted due to oxidative stress caused by a higher selenium concentration (Gupta and Gupta, 2017).

While treatment of real soil leachate using *L. minor* showed the highest Se removal efficiency of about 76%, which was lower than that the Se removal efficiency (86%) by *L. minor* while treatment with artificial wastewater (**Figure 5.3**). Lower removal efficiencies were observed by Miranda et al. (2014), which used *Landoltia punctate* in real mining wastewater instead of standard nutrient solutions. The highest Se removal efficiency was 55% when the wastewater was diluted to 50% and contained 83 μg L^{-1} of Se.

The results of the aquatic plants are similar to the removal of Se by wetland plants that have been used for the remediation of Se wastewater. Johnson et al. (2009) observed up to 70% Se removal efficiency using wild grasses such as bulrush (*Schoenoplectus californicus*) and tamarisk (*Tamarix* sp.). However, in wetlands, most of the Se remains in the sediment and only a small fraction (1%) is taken up by the plants (Johnson et al. 2009). The authors also stated that volatilization contributes to Se removal, in which 17-61% of the selenium is volatilized.

5.4.3. Effect of oxidising agents on phytoremediation of soil leachate

L. minor was more efficient for the removal of Se in most of the experiments and the highest efficiency was obtained when the plants were grown in Hoagland solution, rather than in soil leachate (**Figure 5.1, 5.3**). This can be an indication that the soil leachate may contain elements that are not optimal for the plant growth, such as Cl⁻ and carbonates, or that more nutrients are required for the plants. Also, the growth of the plants in the control samples during the experiments with soil washing were not constant (**Figure 5.4**). When the effect of persulfate was investigated, plants in the control did not grow, as they did in the presence of Mn and SO_4^2. However, there were only small differences in temperature and light during different experiments (**Figure 5.4**), but the plants that were used for the persulfate experiment have been acclimatized with soil leachate (without Se) for a longer period, as all plants were grown in soil washing media before the experiments. This could also explain the fact that *E. densa* had a higher removal efficiency only in this study, as the plant may be more tolerant to the soil leachate. Addition of nutrients can make plants tolerant to heavy metals as they mitigate negative effects in growth (Lebleci and Aksoy 2011).

The effect of soil washing agents on growth of each plant was evident: higher concentrations of the contaminant and chemicals lead to reduction in plant growth (**Figure 5.4**). In contrast, the chlorophyll a content in the plants did not vary considerably. Similar observations were made by Garousi et al. (2015), in which selenite had negative effects in the growth of maize plants when 3 mg kg⁻¹ were added, but there were no significant differences in chlorophyll a and b content, while application of selenate did not have an effect in either growth or chlorophyll content (Garousi et al. 2015). In order to understand this effect of selenium on plant growth and chlorophyll content, Yadav (2010) recommended to study other physiological parameters to have more information about plant stress, such as oxygen radicals (Yadav, 2010). The growth of the plants could also have been affected by the pH of the solution, as this parameter was not controlled and different values were obtained in the experiments, especially for the case of persulfate which decreased the pH to 3. Optimal levels of pH for *L. minor* vary from pH 4-6 (Mclay et al., 1976), while *E. densa* is usually grown at pH 7 (Su et al. 2012). For practical applications, this parameters could be improved by mixing the soil leachate with other water sources, such as agricultural effluents or irrigation water.

Washing the soil with permanganate lead to low concentrations of Mn in solution that plants could uptake and tolerate, while persulfate contributed to SO_4^{2-} which had an inhibitory effect in the Se removal (**Figure 5.3a vs. Figure 5.3e**). Liu et al. (2017) also observed Mn high removal efficiency from *Spirodela pollyrhiza*, which can tolerate Mn at high concentrations of up to 70 mg L⁻¹ and take up 15.75 mg g⁻¹ dry weight. However, in this study a low soil to chemical ratio was used in which KMnO₄ reacted completely. Also, the effects of Mn in the soil need to be further studied to decide if the soil can be reapplied back to the field or if it would need a posttreatment. In the presence of sulfate, the efficiency of plants for selenium removal in the presence of sulfate decreased drastically (**Figure 5.3e**). Bailey et al. (1995) identified that sulfate and selenium have an antagonism as the exposure of aquatic plant *Ruppia maritima* to 13 g L^{-1} of SO$_4$$^{2-}$ and 100 μg L^{-1} of Se decreased the bioconcentration factor of the plant from 1080 to 9.83. However, Saha et al. (2014) used *L. minor* for the treatment of steel wastewater and showed that the plant can remove 30% chloride, 16% sulfate and 14% of the total dissolved solids from wastewater.

The experiment of *L. minor* exposed to a real soil leachate validates the previous experiments (**Figure 5.3**), as the removal efficiencies were in the range found on all the experiments (60-70% removal). The experiment demonstrates that *L. minor* can survive in the soil leachate and that it is effective for Se removal. However, the unaccounted fraction reflects that other mechanisms such as Se volatilization may be involved in the Se removal.

5.5. Conclusion

This study showed that *L. minor* was more efficient for removing Se from seleniferous soil leachates. The presence of Mn in the soil leachate had positive effect on Se removal, while persulfate and sulfate exhibited an inhibitory effect on the Se removal by aquatic plants. This study demonstrated a proof-of-concept on remediation of seleniferous soil by combining soil washing with an environmentally benign soil washing agents and biological treatment of Se-containing leachate. The results demonstrated that phytoremediation using aquatic plants is a sustainable solution for removing Se from soil leachate. Unlike microbial bioremediation, phytoremediation technique does not require addition of carbon or electron donor for removing Se oxyanions. Thus, phytoremediation offers a cost-effective treatment as compared to microbial transformation of Se oxyanions. However, further research is necessary to scale up the process by constructing shallow phytoremediation ponds or a flow through system with two or more ponds suitable for growth of *L. minor* to achieve a lower Se concentration.

References

Bakar, A.F., Yussof, I., Fatt, N., Othman, F., Ashraf, M., 2013. Arsenic, Zinc, and Aluminium Removal from Gold Mine Wastewater Effluents and Accumulation by Submerged Aquatic Plants (*Cabomba piauhyensis, Egeria densa*, and *Hydrilla verticillata*). BioMed Research International, 1-7.

Bailey, F.C., Knight, A.W., Klaine S.J., 1995. Effect of sulfate level on selenium uptake by *Ruppia maritima*. Chemosphere, 30, 579-591.

Bañuelos, G.S., Arroyo, I., Pickering, I.J., Yang, S.I., Freeman, J.L., 2015. Se biofortification of broccoli and carrots grown in soil amended with Se-enriched hyperaccumulator *Stanleya pinnata*. Food Chemistry, 166, 603-608.

Crites, R.W., Middlebrooks, E.J., Bastian, R.K., Reed, S.C., 2014. Natural wastewater treatment systems. 2th edition, CRC Press., Boca Raton, Florida, USA, pp. 248.

Dermont, G., Bergeron, M., Mercier, G., Richer-Laflèche, M., 2008. Soil washing for metal removal: A review of physical/chemical technologies and field applications. J. Hazard. Mater. 152, 1–31.

Dessi, P., Jain, R., Singh, S., Seder-Colomina, M., van Hullebusch, E.D., Rene, E.R., Ahammad, S.Z., Carucci, A., Lens, P.N.L., 2016. Effect of temperature on selenium removal from wastewater by UASB reactors. Water Research, 94, 146-154.

Dhillon, K.S and Dhillon, S.K., 2003. Distribution and management of seleniferous soils. Advances in agronomy, 79, 119-185

Carvalho, K.M and Martin, D.F., 2001. Removal of aqueous Se by four aquatic plants. Journal of Aquatic Plant Management, 39, 33-36.

FAO, WHO, 2001. Human vitamin and mineral requirements. Report of a joint FAO/WHO expert consultation, Bangkok, Thailand.

Garousi, F, Veres, S, Bódi, E., Várallyay, S., Kovács, B., 2015. Role of selenite and selenate uptake by maize. International Journal of Agricultural and Biosystems Engineering, 9, 625-628.

Gupta M., Gupta S., 2017 An overview of selenium uptake, metabolism, and toxicity in plants. Frontiers in Plant Science, 7, 2074.

Hatfield, D.L., Tsuji, P.A., Carlson, B.A., Gladyshev, V.N., 2014. Selenium and selenocysteine roles in cancer, health, and development. Trends in Biochemical Sciences, 39, 112-120.

Handa, N., Bhardwaj, R., Kaur, H., Poonam, Kapoor, D., Rattan, A., Kaur, S., Thukral, A.K., Kaur, S., Arora, S., Kapoor, N., 2015. Se: an antioxidative protectant in plants under stress, In: Ahmad, P. (Ed), Plant metal interaction, Elsevier., India, pp. 179-207.

Haug, A., Graham, R.D., Christophersen, O.A., Lyons, G.H., 2007. How to use the world's scarce Se resources efficiently to increase the Se concentration in food. Microbial Ecology in Health and Disease, 19, 209-228.

Hira, C.K., Partal, K., Dhillon, K.S., 2004. Dietary selenium intake by men and women in high and low selenium areas of Punjab. Public Health Nutrition, 7, 39-43.

Hladun, K.R., Parker, D.R., Trumble, J.T., 2011. Selenium accumulation in the floral tissues of two *Brassicaceae* species and its impact on floral traits and plant performance. Environmental and Experimental Botany, 74, 90-97.

Kamal, M., Ghaly, A.E., Côte, B., 2004. Phytoaccumulation of heavy metals by aquatic plants. Environment International, 29, 1029-1039.

Ko, L., Chang, Y.Y., Lee, C.H., Kim, K.W., 2006. Remediation of soil contaminated with arsenic, zinc, and nickel by pilot-scale soil washing. Environmental Progress, 25(1), 39-48.

Kumari, M., Triphati, B.J., 2014. Effect of aeration and mixed culture of *Eichhornia crassipes* and *Salvinia natans* on removal of wastewater pollutants. Ecological Engineering, 62, 48-53.

Leblebici, Z., Aksoy, A., 2011. Growth and lead accumulation capacity of *Lemna minor* and *Spirodela polyrhiza* (*Lemnaceae*) interactions with nutrient enrichment. Water, air and soil pollution, 214, 175-184.

Lin, H.K., Man, X.D., Walsh, D.E., 2001. Lead removal via soil washing and leaching. The Journal of the Minerals, Metals and Materials Society, 53, 22-25.

Mal, J., Nancharaiah, Y.V., van Hullebusch E.D., Lens, P.N.L., 2016. Effect of heavy metal co-contaminants on selenite bioreduction by anaerobic granular sludge. Bioresource Technology, 206, 1-8.

Marchand, L., Mench, M., Jacob, D.L., Otte, M.L., 2010. Metal and metalloid removal in constructed wetlands, with emphasis on the importance of plants and standardized measurements: a review. Environmental Pollution, 158, 3447-3461.

Mclay C.L., 1976. The effect of pH on the population growth of three species of duckweed: *Spirodela oligorrhiza, Lemna minor* and *Wolffia arrhizal. Fresh Water Biology, 6,125-126.*

Mechora, Š., Cuderman, P., Stibilj, V., Germ, M., 2011. Distribution of Se and its species in *Myriophyllum spicatum and Ceratophyllum demersum* growing in water containing se (VI). Chemosphere 84, 1636-1641.

Mechora, S., Stibilk, V., Germ, M., 2015. Response of *L. minor* to various concentrations of selenite. Environmental Science and Pollution Research, 22, 2216-2224

Megateli, S., Semsari, S., Couderchet, M., 2009. Toxicity and removal of heavy metals (cadmium, copper, and zinc) by *Lemna gibba*. Ecotoxicology and Environmental Safety, 72, 1774-1780

Miranda, A., Muradov, N., Gujar, A., Stevenson, T., Nugedoda, D., Ball, A., Mouradov, A., 2014. Application of aquatic plants for the treatment of selenium-rich mining wastewater and production of renewable fuels and petrochemicals. Journal of Sustainable Bioenergy Systems, 4, 97-112.

Oliveria, H., 2012. Chromium as an environmental pollutant: insights on induced plant toxicity. Journal of Botany, 2012, 1-8.

Ramos, S.J., Faquin, V., Guilherme, L.R.G., Castro, E.M., Ávila, F.W., Carvalho, G.S., Bastos, C.E.A., Oliveria, C., 2010. Selenium biofortification and antioxidant activity in lettuce plants fed with selenate and selenite. Plant, Soil and Environment, 12, 584-588.

Reynier., N., Couder, L., Blais, J.F., Mercier, G., Besner, S., 2015. Treatment of contaminated soil leachate by precipitation, adsorption and ion exchange. Journal of Environmental Chemical Engineering, 3, 977-985.

Rezania, S., Ponraj, M., Talaiekhozani, A., Mohamad, S.E., Md Din, M.F., Taib, S.M., Sabbagh, F., Sairan, F.M., 2015. Perspectives of phytoremediation using water hyacinth for removal of heavy metals, organic and inorganic pollutants in wastewater. Journal of Environmental Management, 163, 125-133.

Saha, P., Shinde, O., Sarkar S. Phytoremediation potential of Duckweed (*Lemna minor L.*) on steel wastewater. International Journal of Phytoremediation.

Schneider, M., Pereira, É.R., Castilho, I.N.B., Carasek, E., Welz, B., Martens, I.B.G., 2016. A simple sample preparation procedure for the fast screening of Se species in soil samples using alkaline extraction and hydride-generation graphite furnace atomic absorption spectrometry. Microchemical Journal, 125, 50-55.

Su, S.Q., Zhou, Y.M., Qin, J.G., Wang, W., Yao, W.Z. & Song, L., 2012. Physiological responses of *Egeria densa* to high ammonium concentration and nitrogen deficiency. Chemosphere, 86, 538-545.

Sunde, R.A., 2012. Se, In: Ross A.C., Caballero, B., Cousins, R.J., Tucker, K.L., Ziegler, T.R, (Eds), Modern nutrition in health and disease, 11th ed, Williams & Wilkins, Philadelphia, USA, pp. 225-237

Sung, M., Lee, C., Lee, S., 2017. Combined mild soil washing and compost-assisted phytoremediation in treatment of silt loams contaminated with copper, nickel, and chromium. Journal of Hazardous Materials, 190, 744-754.

Supriatin, S., Weng, L., Comans, R.N.J., 2015. Se speciation and extractability in Dutch agricultural soils. Science of the Total Environment, 532, 368-382.

USEPA, 2014. External peer review draft. Aquatic life ambient water quality criterion for Se freshwater, United States Environmental Protection Agency (EPA-820-F-14-005).

Van der Spiegel, M., Noordam, M. Y., & van der Fels-Klerx, H. J., 2013. Safety of novel protein sources (Insects, microalgae, aeaweed, duckweed, and rapeseed) and legislative aspects for their application in food and feed production. Comprehensive Reviews in Food Science and Food Safety, 12, 662–678.

Vita, D.P., Platani, C., Fragasso, M., Ficco, D.B.M., Colecchia, S.A., Del Nobile, M.A., Padalino, L., Di Gennaro, S., Petrozza, A., 2017. Se-enriched durum wheat improves the nutritional profile of pasta without altering its organoleptic properties. Food Chemistry, 214, 374-382.

Winkel, L.H.E., Johnson, C.A., Lenz, M., Grundl, T., Leupin, O.X., Amini, M., Charlet, L., 2011. Environmental selenium research from microscopic processes to global understanding.

Wu, S., Wallace, S., Brix, H., Kuschk, P., Kirui, W.K., Masi, F., Dong, R., 2015. Treatment of industrial effluents in constructed wetlands: challenges, operational strategies and overall performance. Environmental Pollution, 201, 107-120.

Wuana, R.A., Okiemen, F.E., 2011. Heavy metals in contaminated soils: a review of sources, chemistry, risks and best available strategies for remediation. International Scholarly Research Network, 2011, 1-20.

Yadav, S, 2010. Heavy metals toxicity in plants: an overview on the role of glutathione and phytochelatins in heavy metal stress tolerance in plants. South African Journal of Botany, 76, 167-179.

Yan, X., Liu, Q., Wang, J., Liao, X., 2017. A combined process coupling phytoremediation and in situ flushing for removal of arsenic in contaminated soil. Journal of Environmental Sciences, 57, 104-109.

Yao, Y., Zhang, M., Tian, Y., Zhao, M., Zhang, B., Zhao, M., Zeng, K., Yin, B., 2017. Duckweed (*Spirodela polyrhiza*) as green manure for increasing yield and reducing nitrogen loss in rice production. Field Crops Research, 214, 272-282.

CHAPTER 6

Selenate reduction by *Delftia lacustris* under aerobic conditions

Abstract

In this research, selenate and selenite reduction by the aerobic bacterium *Delftia lacustris* was investigated and optimised. The selenate reduction profiles of *D. lacustris* were investigated by varying selenate concentration, inoculum size, concentration and source of organic electron donor in minimal salt medium. Although considerable removal of selenate was observed at all concentrations investigated, *D. lacustris* is able to completely reduce 0.1 mM selenate in 96 h using lactate as the carbon source. Around 62.2% unaccounted selenium, 10.9% elemental selenium and 26.9% selenite was observed in the medium after complete reduction of selenate. ^{77}Se NMR spectroscopic analysis showed the unaccounted fraction of selenium was composed of selenium ester compounds. *Delftia lacustris* is reported for the first time as a selenate and selenite reducing bacterium, capable of tolerating and growing in more than 100 mM selenate and 25 mM selenite. Study of enzymatic activity of the cell fractions show that the selenite/selenate reducing enzymes were intracellular and independent of NADPH availability. *D. lacustris* shows unique metabolism of selenium oxyanions to form elemental selenium and water-soluble selenium ester compounds. This novel finding will advance the field of bioremediation of Se-contaminated sites and Se bio-recovery.

Key words: *Delftia lacustris*, aerobic selenate reduction, organic selenium, elemental selenium, selenium esters.

6.1. Introduction

Selenium is an essential trace element in living organisms. It is an important constituent of about 25 known selenoproteins including glutathione peroxidase, thioredoxin reductase, iodothyronine deiodonase and plays an important role in intracellular signaling and redox homeostasis in higher organisms (Huawei, 2009). In spite of its essential requirement and beneficial health effects at low dietary uptake (40 μg d^{-1}), selenium has been listed as a priority pollutant (US EPA) because of its potential bioaccumulation (even at concentrations as low as 5 μg l^{-1}) and associated toxicity (Tan et al., 2016). Selenium toxicity is associated with hair and nail loss and disruption of the nervous and digestive systems in humans and animals (Rayman, 2012; Tinggi, 2008).

Contamination of aquatic bodies with selenium is a serious concern because it can pass through trophic levels, bioaccumulate in living organisms to toxic concentrations and exert toxicity on cellular metabolism (Wu, 2004). Therefore, selenium release through anthropogenic sources such as agricultural drainage water and industrial effluents is tightly regulated. At present, the discharge limit to aquatic bodies has been set at 5 μg l^{-1} Se by the US EPA (Frankenberger Jr. et al., 2004; Zhang et al., 2008). Different methods such as ion exchange, reverse osmosis, adsorption, and chemical reduction using zero valent iron (ZVI) are available for treating selenium containing waters (Frankenberger, Jr. et al., 2004; Sasaki et al., 2008). Among various methods, microbial reduction of soluble selenium oxyanions (selenate and selenite) to insoluble and non-reactive elemental selenium is the best available option because of its cost-effective and eco-friendly nature for treatment of selenium containing wastewaters (Nancharaiah and Lens, 2015a, 2015b).

Reduction of selenium oxyanions to elemental selenium nanospheres has been found in phylogenetically diverse microorganisms isolated from pristine and contaminated environments (Nancharaiah and Lens, 2015a). Selenite reduction has been found in several microorganisms under both aerobic and anaerobic conditions. For example, rhizospheric, nitrogen-fixing bacteria *Azospirullum brasilense* (Tugarova et al., 2014) and *Rhizobium selenireducens* (Hunter *et al.*, 2007) have been reported to reduce selenite to selenium nanoparticles (SeNPs). The Gram positive bacterium *Bacillus megaterium* isolated from mangrove soil (Mishra et al., 2011) and *Bacillus mycoides* isolated from selenium hyper-accumulating plant (Lampis et al., 2014) are able to reduce selenite to elemental selenium. A *Rhodopseudomonas palustris* strain (Li et al., 2014) isolated from a sewage treatment plant and the novel *Paenibacillus selenitireducens* sp. (Yao et al., 2014) isolated from selenium mineral soil collected from a mine in China reduce selenite to elemental selenium anaerobically.

Table 6.1. Overview of bacteria reducing selenate under aerobic conditions

Organism	Optimum Growth Conditions			Se source	Product	Respiration	Reference
	pH	Temp (°C)	Medium				
Pseudomon as stutzeri	7.0	28 °C	Tryptic soy broth	SeO_4^{2-}, SeO_3^{2-}, Se^0	DMSe, DMDSe	Aerobic	Kagami et al. (2013)
NT-I	7–9	20-50 °C	Tryptic soy broth	SeO_4^{2-}, SeO_3^{2-}	Se^0	Aerobic	Kuroda et al. (2011)
Enterobacte r cloacae SLD1a-1	7.2	22 °C	Tryptic soy broth	SeO_4^{2-}	Se^0	Aerobic	Losi & Frankenbe rger Jr. (1997)
Clostridium sp. BXM	7	30 °C for aerobic 37 °C for anaerobic	Basal medium with 10% yeast extract	SeO_4^{2-}, SeO_3^{2-}	Se^0	Aerobic & anaerobic	Bao et al. (2013)
Pseudomon as fluorescens	-	Room temperature	Tryptic soy broth	SeO_4^{2-}, SeO_3^{2-}	Se^0	Aerobic & anaerobic	Hapuarach chi et al. (2004)
Stenotropho monas maltophilia	7.3	Room temperature	Tryptic soy broth	SeO_4^{2-}, SeO_3^{2-}	Se^0, DMSe, DMDSe, DMSeS	Aerobic	Dungan et al. (2003)
Delftia lacustris	7.0	Room temperature	Minimal salt medium with 10 mM lactate	SeO_4^{2-}, SeO_3^{2-}	Se^0, Organic Se	Aerobic	This study

In contrast, only a limited number of selenate reducing bacterial cultures have been isolated and characterized (**Table 6.1**). In the well-studied selenate reducing bacteria *Thauera selenatis* and *Bacillus selenatarsenatis*, the reduction of selenate to selenite is linked to anaerobic respiration and catalysed by selenate reductase (Butler et al., 2012; Yamamura et al., 2007). Very few microorganisms like *Pseudomonas stutzeri* NT-I, *Enterobacter cloacae* SLD1a-1 and *Stenotrophomonas maltophilia*, have been identified to reduce selenate under aerobic conditions (**Table 6.1**), but the understanding of the biochemical mechanism of bacterial selenate reduction under aerobic conditions is rather obscure. *Pseudomonas stutzeri* NT-I is able to convert selenate, selenite and biogenic elemental selenium (Se^0) to volatile selenium compounds (i.e. dimethyldiselenide and dimethylselenide) under aerobic conditions (Kagami et al., 2013). The selenate and selenite reducing bacterium, *Enterobacter cloacae* SLD1a-1 isolated by Losi and Frankenberger Jr. (1997) was hypothesized to reduce selenium oxyanions via membrane-associated reductases, followed by expulsion of the Se^0 precipitate outside the cells. Dungan et al. (2003) speculated detoxification mechanisms for the reduction of both selenate and selenite to elemental Se and volatile Se compounds.

In the present study, a bacterium frequently encountered in a 100 mM selenate stock solution was isolated and identified based on 16S rRNA gene sequencing. The selenate reduction capabilities of this strain were determined by varying the initial selenate concentration, inoculum size, carbon sources and carbon source concentration in minimal salt medium.

6.2. Materials and methods

6.2.1. Chemicals

Sodium selenate (purity > 98%) was procured from Sigma Aldrich. Nutrient broth and agar for isolation and culturing of bacteria was obtained from Oxoid Ltd. (Hampshire, UK). All other analytical-grade chemicals were purchased from Merck. A 100 mM selenate stock was prepared by dissolving 1.89 g of sodium selenate in 100 mL milli-Q water. A sodium lactate (1 M) stock solution was prepared by diluting 13.79 mL of 61.1% sodium lactate to 100 mL with milli-Q water. The stock solutions were stored at 4 °C.

6.2.2. Isolation and growth conditions

The bacterium was isolated serendipitously as a contaminant on nutrient agar plates supplemented with 1 mM selenate from the 100 mM selenate stock solution. Red coloured colonies were re-streaked on mineral salts medium (MSM) supplemented with 1.2% (w/v) agar and 1 mM selenate (**Figure 6.1**). The composition of the MSM was based on the composition of a synthetic wastewater (Stams et al., 1992) (g l^{-1}): 0.053 $Na_2HPO_4.2H_2O$, 0.041 KH_2PO_4, 0.3 NH_4Cl, 0.01 $CaCl_2.2H_2O$, 0.01 $MgCl_2.6H_2O$ and 0.04 $NaHCO_3$ along with 0.1 mL per litre acid and alkaline trace element solutions. The acid trace element stock solution contained (in mM): 7.5 $FeCl_2$, 1 H_2BO_4, 0.5 $ZnCl_2$, 0.1 $CuCl_2$, 0.5 $MnCl_2$, 0.5 $CoCl_2$, 0.1 $NiCl_2$, and 50 HCl. The alkaline trace metals solution was composed of (in mM) 0.1 Na_2WO_4, 0.1 Na_2MoO_4, and 10 NaOH.

MSM agar plates were incubated at 30 °C overnight for the growth of selenate reducing colonies. Individual colonies were transferred to MSM broth containing 1 mM sodium selenate and 10 mM sodium lactate and incubated overnight at 30 °C with constant shaking at 150 rpm. The isolate was sub-cultured in MSM broth and stored in a 50% (v/v) glycerol stock at -20 °C to be used for further experiments for analysing growth and selenate reduction profiles under varying conditions. Liquid samples were collected at regular time intervals for monitoring growth, selenate, selenite, lactate and total selenium.

Figure 6.1. *D. lacustris* cultured on minimal agar plate containing selenate (a) and grown in liquid MSM without (b) and with (c) selenate

6.2.3. Growth and selenate reduction

Growth and selenate reduction under aerobic and anaerobic conditions
Sterile MSM containing 10 mM lactate and 0.1 mM selenate was used to study growth and selenate reduction by the isolate. For anaerobic conditions, the serum bottles were closed with butyl rubber stoppers, crimp sealed and purged with N_2 for 5 min. The bottles containing 100 mL MSM liquid with 0.1 mM selenate were inoculated with an overnight grown bacterial culture under anaerobic conditions. For aerobic growth, the culture flasks plugged with cotton containing 100 mL liquid MSM with 0.1 mM selenate were inoculated with an overnight grown culture and incubated on an orbital shaker set at 150 rpm. Aerobic culture flasks and anaerobic serum bottles were incubated at 30 °C.

Effect of selenate and selenite concentration

Selenate and selenite reduction at different initial selenate and selenite concentrations was determined by inoculating 2% (v/v) overnight grown culture in 100 mL sterile MSM containing 20 mM sodium lactate as carbon source. Growth, selenate and selenite concentrations were determined at different initial selenate and selenite concentrations of 0.1, 0.5, 1 and 2 mM each. Culture flasks were incubated at 30 °C with constant shaking at 150 rpm.

Effect of initial cell density

Initial cell density was measured in terms of colony forming units (CFU) on nutrient agar plates per millilitre of samples. The cell density was varied by inoculating 1%, 2% or a cell pellet of an overnight grown culture (5.4×10^{12} CFU mL^{-1}) in sterile MSM containing 10 mM sodium lactate and 0.1 mM sodium selenate, incubated at 30 °C with constant shaking at 150 rpm.

Effect of carbon source and concentration

Growth and selenate reduction was determined in the presence of different carbon sources such as lactate, acetate and glucose. Tests were carried out by inoculating 2% (v/v) overnight grown culture in sterile MSM each containing 0.1 mM sodium selenate and 20 mM sodium lactate, 20 mM sodium acetate, 20 mM sodium citrate, 20 mM mannitol or 55 mM D-glucose, incubated at 30 °C with constant shaking at 150 rpm. Growth and selenate reduction were also determined at different initial lactate concentrations of 10 mM and 20 mM. A fixed selenate concentration of 0.1 mM sodium selenate was used. Culture flasks were incubated at 30 °C with constant shaking at 150 rpm.

Effect of tungstate on selenate and selenite reduction

The effect of tungstate on selenate and selenite reduction was determined by inoculating 2% of fully grown culture in sterile MSM containing 0.1 mM sodium selenate, 10 mM sodium tungstate and 20 mM lactate. MSM containing 0.1 mM sodium selenate and 20 mM lactate inoculated with bacterial cultures was used as control. For determining the effect of tungstate on selenite reduction, a similar experiment was performed by replacing selenate with selenite.

6.2.4. Selenate reduction by spent medium, cell lysate and resting cells

The bacterial culture was grown in nutrient broth overnight. This culture was centrifuged at 5000 g for 15 min to pellet the bacterial cells. The supernatant was filter-sterilized with 0.22 μm sterile filter disks. Control samples were obtained after heat inactivation by boiling at 121 °C for 15 min. Sodium selenate was added to filtered spent medium at a final concentration of

0.1 mM and incubated at 30 °C. Selenate and selenite concentrations were analysed at 0, 2, 4, 8, 24, 48 and 72 h of incubation. The pellet was re-suspended and washed twice with 10 mM Tris HCl, pH 7.5. Bacterial cells were lysed by pulse sonication at 4 °C. Sodium selenate or sodium selenite were added separately to make the final concentration to 0.1 mM and incubated at 30 °C. Selenate and selenite were analysed at 0 and 4 h of incubation. The effect of NADPH addition was analysed by adding 2 mM NADPH to the lysed cells and supernatant along with sodium selenate and selenite separately.

To obtain the resting cells, the strain was grown at 30 °C in nutrient broth for 48 h to ensure complete utilization of the energy source in the medium and microbial cells were in the stationary phase. The cells were then harvested by centrifugation at 5000 rpm for 10 min, followed by washing and inoculation in sterile MSM. Sterile sodium selenate stock was added to MSM containing resting cells to make the final concentration of 0.1 mM and incubated at 30 °C with constant shaking at 150 rpm under aerobic or anaerobic conditions. Selenate, selenite and total selenium were analysed at regular time intervals.

6.2.5. Minimum inhibitory concentration of selenate

Nutrient broth and MSM containing a final concentration of 0, 0.78, 1.56, 3.12, 6.25, 12.5, 25, 50, and 100 mM of sodium selenate and 10 mM lactate as carbon source with a final volume of 100 mL was sterilized by autoclaving. The flasks were inoculated with 1% bacterial culture $(3.25 \times 10^{12}$ CFU mL^{-1}) so that the initial absorbance at 600 nm (indicating cell density) was around 0.02 and incubated at 30 °C with constant shaking at 150 rpm overnight. Growth was monitored by measuring absorbance at 600 nm. The concentration which caused complete inhibition in bacterial growth was described as the minimum inhibitory concentration.

6.2.6. Extraction of organic selenium compound from spent culture medium

Culture was grown in selenate or selenite containing MSM as described above. After 5 days of incubation, the culture was centrifuged at 4000 rpm for 20 min to separate cells from the spent MSM and the supernatants were extracted by 2-phase liquid-liquid extraction using ethyl acetate. The aqueous and ethyl acetate phases were analysed by thin layer chromatography (TLC) as described by Bottura and Pavesi (1987). The solvent system used was ethyl acetate: methanol (4:1) and after the run on the silica gel TLC plates (Sigma Aldrich, USA), the spots were observed under UV light. The ethyl acetate phase was further concentrated in a rotary evaporator. A white precipitate formed at the bottom of the flask was then dissolved in dimethyl sulfoxide (DMSO) for ^{77}Se NMR analysis on a Bruker Avance 500 MHz (Rheinstetten, Germany). The NMR were recorded at the frequency of 95.41 MHz using deuterated DMSO as a solvent.

6.2.7. 16S rRNA gene sequencing

The genomic DNA of the bacterial isolate was extracted using the PowerSoil DNA isolation kit (MO BIO Laboratories Inc., USA). For this, the strain was grown overnight in nutrient broth and centrifuged at 4000 rpm for 10 min. The genomic DNA was isolated from the cell pellet by following the manufacturer's instructions. The 16S rRNA gene was PCR amplified from genomic DNA using gene specific primers and analysed by DNA sequencing by Eurofins Genomics (Ebersberg, Germany). The sequence was blasted in the nucleotide database of the US *National Center for Biotechnology Information* (NCBI, www.ncbi.nlm.nih.gov) to obtain the phylogenetic identity of the isolate.

6.2.8. Electron microscopic imaging

The morphology of the selenate reducing bacterial cells was analysed using a Jeol JSM-6010LA scanning electron microscopy (SEM) (Nieuw Vennep, Netherlands). The culture was washed with phosphate buffered saline, pH 7, followed by fixation with 2.5% glutaraladehyde and dehydration using a graded ethanol series (Kaláb *et al.*, 2008). Bacterial cells grown in nutrient broth, and minimal salt medium with and without 0.1 mM selenate were fixed in glutaradehyde and dehydrated as described by Kaláb *et al.* (2008) and subsequently observed using SEM.

6.2.9. Analysis

Growth was routinely monitored by measuring absorbance at 600 nm using UV-Vis spectrophotometry (Shimadzu UV-2501 PC, Groningen, Netherlands). The samples were filtered through a 0.22 μm syringe filter and analysed for selenate, selenite and lactate. Selenite was analysed by UV-Vis spectrophotometry using ascorbic acid reagent (Mal et al., 2016). Selenate was measured by an ion chromatograph equipped with AS14A 3 mm analytical column (Dionex ICS series AS-DV, USA) using 8 mM sodium carbonate and 1 mM sodium bicarbonate as eluant with a flow rate of 0.5 mL min^{-1} (Dessì et al., 2016). The retention time for selenate was 8.5 min.

After acidification of samples using 65% nitric acid, total selenium was analysed by a graphite furnace atomic absorption spectrophotometer (GF-AAS) (Solaar AA Spectrometer GF95, Thermo Electrical, Unity lab services, England) as described by Dessì *et al.* (2016). For elemental Se quantification, the samples were centrifuged at 37,000 x g for 15 min to separate Se(0) particles and bacterial cells from dissolved Se (Mal et al., 2016). Pellet and supernatant samples were then digested with 65% nitric acid, diluted and analysed by GF-AAS as intracellular and extracellular elemental Se particles and total dissolved Se, respectively.

Unaccounted selenium was calculated by subtracting the selenate and selenite concentration from the total dissolved selenium:

$$\text{Unaccounted Se} = \text{Total dissolved Se} - (\text{Selenate} + \text{Selenite}) \qquad \text{(Eq. 6.1)}$$

Where total dissolved Se, selenate and selenite were measured using GF-AAS, ion chromatography and UV-Vis spectrophotometry, respectively, as described above.

6.3. Results

6.3.1. Electron microscopic imaging and identification by 16S rRNA gene sequencing

The 16 rRNA gene sequence of the selenate reducing isolate showed 100% homology with *Delftia lacustris*. Selenate reduction of *D. lacustris* under aerobic growth conditions was intriguing (**Figure 6.1**), hence the reduction process was investigated in detail.

Figure 6.2. Scanning electron microscopic images of *D. lacustris* showing distinct differences in the initial inoculum cultured in nutrient broth (a) and micro-organisms cultured in minimal salt medium (b and c). However, the presence (c) or absence (b) of selenate does not visually affect bacterial morphology in MSM

The scanning electron microscopic images showed distinct variation in the cell morphology of the *D. lacustris* cells, with rod shaped cells in chains when cultured in nutrient broth medium and rod shaped bacterial cells in the form of aggregates when cultured in minimal medium (**Figure 6.2**). However, no significant difference was observed in the morphology of bacterial cultures grown in MSM with or without 0.1 mM selenate.

6.3.2. Growth and selenate reduction profiles

Lactate and selenate removal under aerobic and anaerobic conditions

Lactate and selenate removal profiles under aerobic and anaerobic conditions are shown in **Figure 6.3**. In aerobic conditions, lactate was completely utilized within the first 10 h (**Figure 6.3a**). Selenate removal was around 50% within 10 h of inoculation. However, no major change in either lactate or selenate concentration was observed under anaerobic conditions (**Figure 6.3b**). There was no change in the turbidity, suggesting that *D. lacustris* is incapable of growth (Data not shown), lactate consumption or selenate reduction (**Figure 6.3b**) under anaerobic conditions.

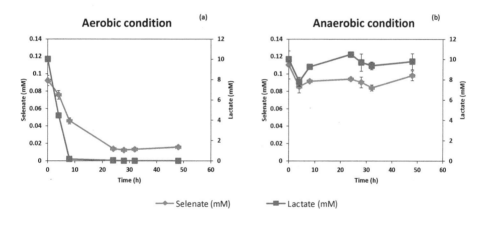

Figure 6.3. Selenate and lactate removal profiles by *D. lacustris* in mineral salts medium under aerobic (a) and anaerobic (b) growth conditions

Carbon source type and concentration and initial cell density

D. lacustris was able to grow by utilizing different carbon sources such as acetate, citrate, lactate and mannitol under aerobic condition (**Figure 6.4a**). However, it was unable to use glucose as the carbon source. Moreover, selenate was removed from the MSM when supplemented with acetate, citrate, lactate and mannitol, but not glucose as the carbon source. Reduction of selenate by *D. lacustris* was more efficient when lactate, acetate and mannitol were the carbon sources (**Figure 6.4a**). Se speciation showed formation of selenite, Se(0) and unaccounted dissolved Se after 7 d of incubation (**Figure 6.4b**).

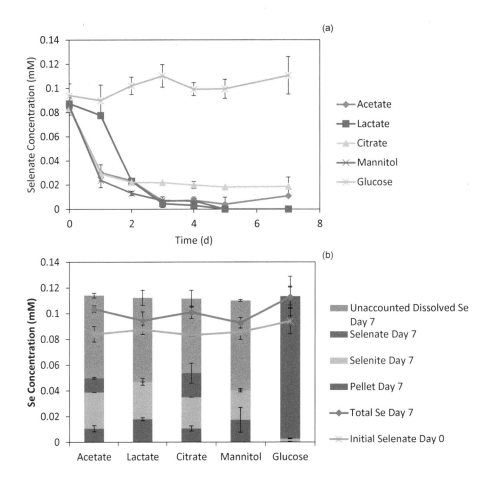

Figure 6.4. Selenate reduction (a) by *D. lacustris* in mineral salts medium supplemented with different carbon sources and selenium speciation (b) during selenate reduction by *D. lacustris* in the presence of different carbon sources

A major portion of the dissolved Se which could not be accounted for neither as selenate nor selenite and is furthur referred to as unaccounted dissolved Se (**Figure 6.4b**). Reduction of selenate and growth of the bacterium was increased with increasing concentration of initial lactate concentration. In MSM containing 20 mM lactate, complete reduction of 0.1 mM selenate was observed after 96 h (**Figure 6.5**). Selenate removal profiles typically consisted of an initial lag phase, rapid removal phase and a stable phase without selenate removal (**Figure 6.5**).

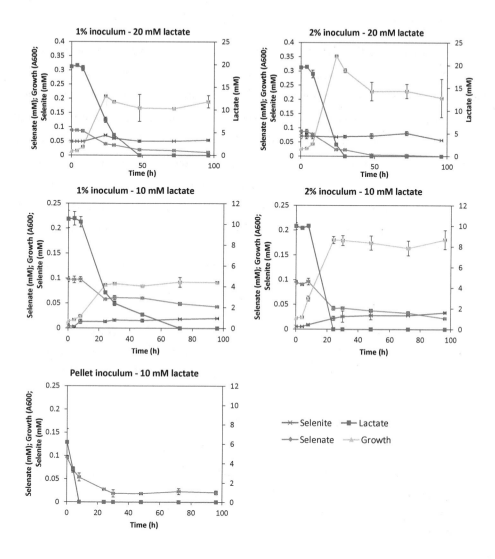

Figure 6.5. Selenate reduction by *D. lacustris* in mineral medium inoculated with 1 and 2% cell densities at different lactate concentrations

Rapid reduction of selenate was observed when the initial cell density of *D. lacustris* inoculated in the medium was higher (**Figure 6.5**). Selenate reduction was observed by *D. lacustris* at two inoculum densities. Cell pellets harvested from the overnight grown *D. lacustris* culture were resuspended and added to medium at 1 and 2% (v/v). Complete reduction of 0.1 mM selenate was observed after 96 h when the 2% inoculum was added to the culture (**Figure 6.5**). The selenate reduction rate was unchanged with increase in inoculum concentration beyond 2% of its initial concentration.

Selenate and selenite concentration

Complete selenate reduction was observed only when the selenate concentration in the medium was 0.1 mM. The percentage reduction decreased with increased selenate concentration in the medium. In contrast, reduction of selenite was incomplete even at the lowest (0.1 mM) selenite concentration tested. Selenite was reduced by 60-72% in all the culture flasks, irrespective of the initial selenite concentration in the medium (**Table 6.2**). Se speciation analysis showed no major change in elemental Se and unaccounted Se ratio during transformation of different selenate concentrations by *D. lacustris*. No major change in the elemental Se concentration was noticed during transformation of different selenite concentrations. However, the fraction of unaccounted Se increased with increasing initial selenite concentration (**Figure 6.6**).

Table 6.2. Removal efficiencies and concentrations of elemental selenium and unaccounted selenium produced by *D. lacustris* after 96 h incubation at 30 °C with varying initial concentration of selenate and selenite

Se source	Selenate			Selenite		
Concentration (mM)	Removal efficiency (%)	Elemental Se (mM)	Unaccounted Se* (mM)	Removal efficiency (%)	Elemental Se (mM)	Unaccounted Se* (mM)
0.1	100	0.009 ± 0.002	0.051 ± 0.009	69.5 ±0.8	0.022 ± 0.002	0.040 ± 0.002
0.2	71.6 ±3.3	0.003 ± 0.0003	0.095 ± 0.007	70.9 ±2.6	0.031 ± 0.009	0.085 ± 0.009
0.3	54.1 ±9.9	0.006 ± 0.001	0.102 ± 0.015	71.9 ±0.3	0.038 ± 0.003	0.148 ± 0.002
0.4	49.1 ±0.7	0.007 ± 0.002	0.120 ± 0.005	69.8 ±0.5	0.045 ± 0.001	0.168 ± 0.001
0.5	42.9 ±2.7	0.008 ± 0.001	0.113 ± 0.003	60.2 ±2.4	0.043 ± 0.006	0.215 ±0.001

*Unaccounted Se was determined by subtracting dissolved selenate and selenite from total dissolved selenium after 96 h.

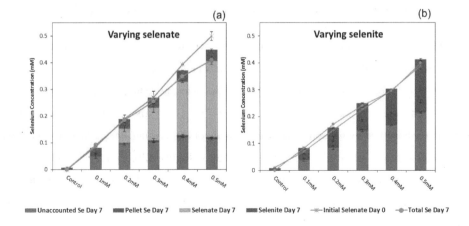

Figure 6.6. Selenate (a) and selenite (b) reduction at varying concentrations from 0.1 - 0.5 mM showing reduction of selenate and selenite to elemental Se, and an unidentified soluble organo-Se compound. Pellet Se represents elemental selenium as well as organic selenium present intracellularly

Minimum inhibitory concentration of selenate

Bacterial growth was observed in all the tested concentrations up to 100 mM selenite in both nutrient broth as well as MSM containing lactate (**Figure 6.7**). Growth of *D. lacustris* was only slightly inhibited by 50 and 100 mM selenate in nutrient broth. In contrast, the growth in mineral medium was decreased in presence of all the selenate concentrations tested. The strain was, nevertheless, able to grow even in the presence of 100 mM selenate in this medium. Thus, the minimum inhibitory concentration of selenate was >100 mM.

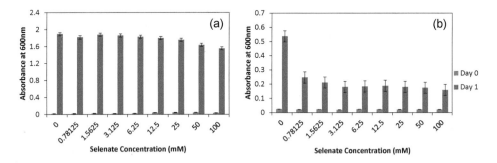

Figure 6.7. *D. lacustris* growth in the presence of different concentrations of selenate in nutrient broth (a) and minimal salt medium (b)

6.3.3. Selenate reduction by spent growth medium, cell lysate and resting cells

Selenate reduction did not occur in spent growth medium without *D. lacustris* cells. The selenate reduction ability was lost when *D. lacustris* cells were lysed or autoclaved (**Figure 6.8**). No significant difference was observed in the selenate reduction profile between the untreated and sterile lysate sample and control Tris HCl. About 60% of selenite was immediately adsorbed by the cell lysate with or without autoclaving. The reduction continued by the cell lysate and removed up to 80% after 6 hours (**Figure 6.8**). Untreated cell lysate was shown to reduce selenite, whereas autoclaved lysate and control Tris HCl did not induce selenite reduction. The autoclaved lysate did not show selenite reduction, suggesting the heat labile nature of the reducing agent or reductase enzyme.

No reduction or adsorption of selenate or selenite was observed in control setups with Tris HCl. Resting cells also did not reduce selenate under both aerobic and anaerobic conditions. Addition of NADPH did also not increase the reduction rate of selenate or selenite (Data not shown) by the lysed cells or the supernatant, indicating the selenate-reducing enzymes are NADPH independent.

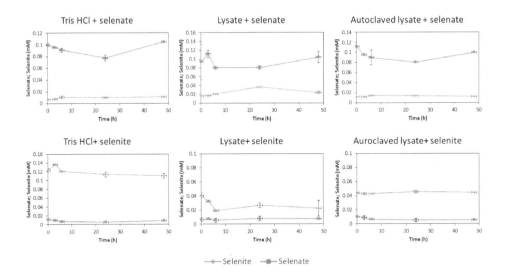

Figure 6.8. Selenite and selenate reduction by *D. lacustris* cell lysate

6.3.4. Effect of tungstate on selenate and selenite reduction

In selenate reduction experiments, growth and selenate reduction were unaffected by the presence of sodium tungstate in the medium. However, the selenite concentration in the medium could not be measured because sodium tungstate interfered with the selenite measurement by the UV-Vis spectrophotometric method. Visual inspection showed nevertheless that the characteristic red colour of elemental Se did not develop in the medium containing tungstate (**Figure 6.9**).

Figure 6.9. *D. lacustris* produces characteristic red elemental Se in the medium containing selenite with lactate (right flask), whereas in the presence of tungstate (left flask) the culture is colourless suggesting inhibition of elemental Se production by tungstate

This suggests that reduction of selenite to elemental Se by *D. lacustris* was inhibited by the presence of sodium tungstate. Tungstate has been reported as a competitive inhibitor of molybdenum of the membrane-bound molybdo-enzyme active site, a putative selenate reductase in *E. cloacae* SLD1a-1 (Watts et al., 2003).

6.3.5. Organic selenium analysis

TLC was performed to determine the components present in aqueous extract as well as ethyl acetate extracts obtained from both selenate and selenite-reducing culture medium. **Table 6.3** defines each spot and the Rf values observed in the TLC plates of aqueous and ethyl acetate extracts. Two spots each were observed for samples A, B and D (**Figure 6.10**), with the Rf value of the first spot matching with that of the standard (Selenium dioxide standard solution). When observed under UV light, the spot glowed indicating the presence of conjugated double Se bonds of an aromatic compound, like those present in phenyl selenide (PhSe) compounds or selenate esters. The standard indicated (**Table 6.3**) the presence of Se compounds in the extracts (Rf 0.33). The spots with higher Rf values also exhibited fluorescence under UV indicating the presence of aromatic compounds.

Table 6.3. Spot description and Rf values observed in the thin layer chromatography. Each Rf value represent different spots observed on the TLC plate

Spots	Description	Rf values
A	Aqueous extract for bacteria growing in selenate in MSM	0.33, 0.93
B	Ethyl acetate extract for bacteria growing in selenate in MSM	0.33, 0.87
C	Aqueous extract for bacteria growing in selenite in MSM	0.3
D	Ethyl acetate extract for bacteria growing in selenite in MSM	0.37, 0.87
E	Selenium dioxide standard solution (Se reference)	0.33

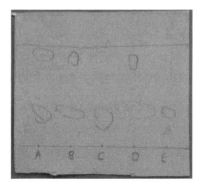

Figure 6.10. TLC of aqueous and ethyl acetate extracts was carried out as a primary method to separate and identify the mixture. Spots A and C are aqueous extract from selenate and selnite containing spent medium respectively. Spot B and D are ethyl acetate extract from selenate and selnite containing spent medium respectively. Spot E is 2% selnium dioxide solution dissolved in nitric acid (Se reference).

The NMR spectra obtained after analysing the samples extracted with ethyl acetate showed peaks for [77]Se at around 1910.15 ppm, which correspond to selenium ester compounds based on the ^{77}Se NMR databank (**Figure 6.11**). However, the peak had a low intensity indicating a rather low concentration of the selenium compound in the extract investigated. Ion trap mass spectrometry and Fourier Transform Ion Cyclotron Resonance Mass Spectrometry (FT-ICR MS) analysis of the extracts could not further confirm the chemical structure of the organic selenium compound produced by *D. lacustris* (Data not shown).

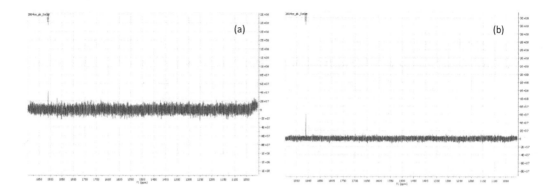

Figure 6.11. NMR spectra for the ethyl acetate extract of the MSM broth after reduction of selenate (a) and selenite (b) by *D. lacustris*

6.4. Discussion

6.4.1. Selenate reduction by *D. lacustris* is linked to growth and availability of carbon source

D. lacustris was first isolated from mesotrophic lake water in 2009 as a peptidoglycan degrading bacterium with extracellular lysozyme and chitinase activity (Jørgensen et al., 2009). The bacterium is aerobic, Gram negative, non-spore forming, motile, rod shaped and incapable of utilizing glucose as the carbon source. The strain was found to be oxidase and catalase positive and utilizes oxygen as terminal electron acceptor. *D. lacustris* has been reported to reduce nitrate to nitrite (Jørgensen et al., 2009), but there are no reports on its ability to reduce either selenate or selenite. This is the first report on the reduction of selenate and selenite by *D. lacustris*.

D. lacustris isolated from a selenate stock solution showed consistent reduction of selenate under aerobic growth conditions (**Figure 6.3**). The selenate reduction ability depended on the inoculum density (**Figure 6.5**), carbon source type (**Figure 6.4**) and concentration (**Figure 6.5**). Rapid growth and selenate reduction was observed when grown in the presence of higher lactate concentrations (**Figure 6.5**). The selenate reduction ability of the strain was associated with carbon source availability and cell growth. Interestingly, selenate reduction was not observed in the resting cell experiment, suggesting that selenate reduction by this strain only occurs in the presence of external carbon source and is linked to growth.

The selenite reduction efficiency was inhibited in the presence of tungstate (**Figure 6.9**), suggesting that the selenite reduction by *D. lacustris* is molybdenum-dependent. In contrast, selenate reduction was unaffected by tungstate (Data not shown) suggesting the selenate reducing by *D. lacustris* is molybdenum-independent. This shows that different enzymes or molecules are involved in selenate and selenite reduction. Similarly, Zheng *et al.* (2014) inferred that different enzymes were involved in selenate and selenite reduction in *Comamonas testosterone*, where selenate reduction by the microorganism is a molybdenum dependent reduction inhibited by tungstate, while NADPH dependent enzymes are involved in selenite reduction. In this study, however, reduction of both selenate and selenite by *D. lacustris* was found to be NADPH independent. Further characterisation of the enzymology and intermediates of the selenate metabolism of *D. lacustris* is required. Nevertheless, attributes like fewer growth requirements, ease of culturing and selenate reduction under aerobic conditions deems *D. lacustris* as a potential microorganism for bioremediation of selenate bearing wastewaters and soils.

6.4.2. Reduction of selenate to elemental Se and unaccounted soluble Se fraction

In the experiment with varying selenate and selenite concentrations, selenate removal accompanied by the appearance of selenite and elemental Se (**Figure 6.6**) confirms reduction of selenate by *D. lacustris* under aerobic conditions. The negligible difference between the initial and final total selenium concentration (**Figure 6.6**) suggested an insignificant role of volatilization in selenate or selenite removal by *D. lacustris*. Upon subtracting the selenate and selenite concentrations from total dissolved Se at the end of the experiment, a significant fraction of unaccounted form of Se was observed in dissolved form, which could be in the form of assimilatory Se compounds such as alkyl selenide or other organo-Se compounds. Se speciation analysis found that more than 60% of the selenate was directed towards organo-Se compound synthesis, irrespective of the initial selenate concentration (**Table 6.2**). Similar experiments performed with variable initial selenite concentration showed that the concentration of the organo-Se compound was increased at the end of the experiment by almost 20% with respect to the percentage of total selenite removed when increasing the initial selenite concentration from 0.1 mM to 0.5 mM (**Table 6.2**).

After 7 days of incubation, the amount of elemental Se was found to be decreasing with increasing of initial selenate or selenite concentration. Thus, selenate and selenite reduction was mainly directed towards the formation of organo-Se compounds. Similar results were observed for a *Ralstonia metallidurans* strain, where selenite was reduced to alkyl selenide by an assimilatory pathway, followed by detoxification to form Se(0) (Sarret et al., 2005). However, ion trap mass spectrometry and FT-ICR MS analysis of the ethyl acetate extracts from the supernatant of the *D. lacustris* cultures grown in selenate and selenite did not identify alkyl selenium as an accumulated intermediate (Data not shown).

Selenate reduction by *Thauera selenatis* is catalysed by intracellular thiol glutathione and (Debieux et al., 2011) suggested that there is a link between reductive and volatile methylated selenide dependent detoxification. Other enzymes known to be catalysing selenite reduction are nitrate reductase (Butler et al., 2012; Hunter, 2014), coupling with sulfate reduction (Hockin and Gadd, 2003) or by a glutathione reductase enzyme (Hunter and Manter, 2011; Kessi and Hanselmann, 2004). The role of the thioredoxin and thioreductase enzymes (Garbisu et al., 1999), chromate reductase (Rath et al., 2014) and NADH dependent reductase activity (Dwivedi et al., 2013; Hunter, 2014) in selenite reduction has been widely studied. Selenite reduction in different microorganisms occurs by different enzymes present either in the cytoplasm (Antonioli et al., 2007; Hunter, 2014; Zheng et al., 2014) or the periplasm (Demoll-decker, 1993). Lampis *et al.* (2014) reported that the selenite reduction activity of *Bacillus mycoides* is linked to the initial selenite concentration and total number of bacterial cells, rather than the bacterial growth phase. The selenite reduction profile of *B. mycoides*, when carried out under aerobic conditions with increasing initial selenite concentration, shows that the selenite reduction rate increases with increased selenite concentration (**Figure 6.6b**)

6.4.3. Intracellular enzymes are involved in selenite and selenate reduction

D. lacustris is known to produce extracellular enzymes, i.e. chitinase and lysozyme and has been the first microorganism to report for extracellular degradation of peptidoglycan (Jørgensen et al., 2009). However, in the present study, experiments with microbial cell lysate suggest that the microorganism reduces selenite and selenate intracellularly to organo-Se compounds and red elemental Se. The enzyme involved in selenate and selenite reduction to red elemental-Se is a molybdenum dependent intracellular enzyme (**Figure 6.9**).

Extracellular reduction of selenite has been reported previously by Lampis *et al.* (2014). Selenite was reduced by cell lysate of *D. lacustris* under ambient conditions (**Figure 6.8**), suggesting that selenite reducing enzymes or molecules are produced constitutively by the bacterium. In contrast, selenate reduction was not achieved using cell lysate (**Figure 6.8**), probably because it is an energy-driven reaction necessitating reducing equivalents or other co-factors. Previous studies have shown that supplementation of NADPH and NADH to the cell lysate and supernatant has improved reduction of selenite under abiotic conditions (Zheng et al., 2014).

6.5. Conclusion

Removal of selenate is possibly mediated by two parallel mechanisms which form elemental Se and an unidentified non-volatile Se compound, likely a selenite ester. TLC and NMR spectroscopy suggest that the unidentified Se is an organic-Se compound which accumulated in the growth medium during selenate or selenite removal by *D. lacustris*. The observed shift in Se signal of the NMR spectra is indicative of the chemical environment of the Se atom in the molecule. Based on ^{77}Se NMR databank analysis, this shift is similar to the one obtained for compounds such as selenite and selenate esters. To our knowledge, there are no reports on the formation of non-volatile selenate esters during microbial transformation of Se oxyanions. The ^{77}Se shift, however, alone cannot give information about the exact chemical identity of the selenium compound. Further characterization using ^{1}H and ^{13}C NMR, mass spectrometry and elemental composition analysis is required to obtain the chemical structure of the organo-Se compound formed by *D. lacustris*.

References

Antonioli, P., Lampis, S., Chesini, I., Vallini, G., Rinalducci, S., Zolla, L., Righetti, P.G., 2007. *Stenotrophomonas maltophilia* SeITE02, a new bacterial strain suitable for bioremediation of selenite-contaminated environmental matrices. Appl. Environ. Microbiol. 73, 6854–6863.

Bao, P., Huang, H., Hu, Z.Y., Häggblom, M.M., Zhu, Y.G., 2013. Impact of temperature, CO_2 fixation and nitrate reduction on selenium reduction, by a paddy soil *Clostridium* strain. J. Appl. Microbiol. 114, 703–712.

Bottura, G., Pavesi, M.A., 1987. Thin-layer chromatography of some organic selenium compounds and their oxygen and sulfur analogs. Microchem. J. 35, 112–119.

Butler, C.S., Debieux, C.M., Dridge, E.J., Splatt, P., Wright, M., 2012. Biomineralization of selenium by the selenate-respiring bacterium *Thauera selenatis*. Biochem. Soc. Trans. 40, 1239–43.

Debieux, C.M., Dridge, E.J., Mueller, C.M., Splatt, P., Paszkiewicz, K., Knight, I., Florance, H., Love, J., Titball, R.W., Lewis, R.J., Richardson, D.J., Butler, C.S., 2011. A bacterial process for selenium nanosphere assembly. Proc. Natl. Acad. Sci. U. S. A. 108, 13480–5.

Demoll-decker, H., 1993. The periplasmic nitrate reductase of *Thauera selenatis* may catalyse the reduction of selenite and elemental selenium. Arch. Microbiol. 160, 241–247.

Dessi, P., Jain, R., Singh, S., Seder-Colomina, M., van Hullebusch, E.D., Rene, E.R., Ahammad, S.Z., Carucci, A., Lens, P.N.L., 2016. Effect of temperature on selenium removal from wastewater by UASB reactors. Water Res. 94, 146–154.

Dungan, R.S., Yates, S.R., Frankenberger, W.T., 2003. Transformations of selenate and selenite by *Stenotrophomonas maltophilia* isolated from a seleniferous agricultural pond sediment. Environ. Microbiol. 5, 287–295.

Dwivedi, S., AlKhedhairy, A.A., Ahamed, M., Musarrat, J., 2013. Biomimetic synthesis of selenium nanospheres by bacterial strain JS-11 and its role as a biosensor for nanotoxicity assessment: A novel Se-bioassay. PLoS One 8, e57402.

Frankenberger, Jr., W.T., Amrhein, C., Fan, T.W.M., Flaschi, D., Glater, J., Kartinen, Jr., E., Kovac, K., Lee, E., Ohlendorf, H.M., Owens, L., Terry, N., Toto, A., 2004. Advanced treatment technologies in the remediation of seleniferous drainage waters and sediments. Irrig. Drain. Syst. 18, 19–42.

Garbisu, C., Carlson, D., Adamkiewicz, M., Yee, B.C., Wong, J.H., Resto, E., Leighton, T., Buchanan, B.B., 1999. Morphological and biochemical responses of *Bacillus subtilis* to selenite stress. BioFactors 10, 311–319.

Hapuarachchi, S., Swearingen, J., Chasteen, T.G., 2004. Determination of elemental and precipitated selenium production by a facultative anaerobe grown under sequential anaerobic/aerobic conditions. Process Biochem. 39, 1607–1613.

Hockin, S.L., Gadd, G.M., 2003. Linked redox precipitation of sulfur and selenium under anaerobic conditions by sulfate-reducing bacterial biofilms. Appl. Enivironmental Microbiol. 69, 7063–7072.

Huawei, Z., 2009. Selenium as an essential micronutrient: Roles in cell cycle and apoptosis. Molecules 14, 1263–1278.

Hunter, W.J., 2014. A *Rhizobium selenitireducens* protein showing selenite reductase activity. Curr. Microbiol. 68, 311–6.

Hunter, W.J., Kuykendall, L.D., Manter, D.K., 2007. *Rhizobium selenireducens* sp. nov.: a selenite-reducing *α-Proteobacteria* isolated from a bioreactor. Curr. Microbiol. 55, 455–60.

Hunter, W.J., Manter, D.K., 2011. *Pseudomonas seleniipraecipitatus* sp. nov.: A selenite reducing *γ-Proteobacteria* isolated from soil. Curr. Microbiol. 62, 565–569.

Jørgensen, N.O.G., Brandt, K.K., Nybroe, O., Hansen, M., 2009. *Delftia lacustris* sp. nov., a peptidoglycan-degrading bacterium from fresh water, and emended description of *Delftia tsuruhatensis* as a peptidoglycan-degrading bacterium. Int. J. Syst. Evol. Microbiol. 59, 2195–2199.

Kagami, T., Narita, T., Kuroda, M., Notaguchi, E., Yamashita, M., Sei, K., Soda, S., Ike, M., 2013. Effective selenium volatilization under aerobic conditions and recovery from the aqueous phase by Pseudomonas stutzeri NT-I. Water Res. 47, 1361–1368.

Kaláb, M., Yang, A., Chabot, D., 2008. Conventional scanning electron microscopy of bacteria. Infocus 44–61.

Kessi, J., Hanselmann, K.W., 2004. Similarities between the abiotic reduction of selenite with glutathione and the dissimilatory reaction mediated by *Rhodospirillum rubrum* and *Escherichia coli*. J. Biol. Chem. 279, 50662–50669.

Kuroda, M., Notaguchi, E., Sato, A., Yoshioka, M., Hasegawa, A., Kagami, T., Narita, T., Yamashita, M., Sei, K., Soda, S., Ike, M., 2011. Characterization of Pseudomonas stutzeri NT-I capable of removing soluble selenium from the aqueous phase under aerobic conditions. J. Biosci. Bioeng. 112, 259–264.

Lampis, S., Zonaro, E., Bertolini, C., Bernardi, P., Butler, C.S., Vallini, G., 2014. Delayed formation of zero-valent selenium nanoparticles by *Bacillus mycoides* SeITE01 as a consequence of selenite reduction under aerobic conditions. Microb. Cell Fact. 13, 35.

Li, B., Liu, N., Li, Y., Jing, W., Fan, J., Li, D., Zhang, L., Zhang, X., Zhang, Z., Wang, L., 2014. Reduction of selenite to red elemental selenium by *Rhodopseudomonas palustris* strain N. PLoS One 9, e95955.

Losi, M.E., Frankenberger Jr., W.T., 1997. Reduction of selenium oxanions by *Enterobacter cloacae* strain SLD1a-1: Isolation and growth of the bacterium and its expulsion of selenium particles. Appl. Enivironmental Microbiol. 16, 3079–3084.

Mal, J., Nancharaiah, Y. V., van Hullebusch, E.D., Lens, P.N.L., 2016. Effect of heavy metal co-contaminants on selenite bioreduction by anaerobic granular sludge. Bioresour. Technol. 206, 1–8.

Mishra, R.R., Prajapati, S., Das, J., Dangar, T.K., Das, N., Thatoi, H., 2011. Reduction of selenite to red elemental selenium by moderately halotolerant *Bacillus megaterium* strains isolated from Bhitarkanika mangrove soil and characterization of reduced product. Chemosphere 84, 1231–7.

Nancharaiah, Y. V, Lens, P.N.L., 2015a. The ecology and biotechnology of selenium respiring bacteria. Microbiol. Mol. Biol. Rev. 79, 61–80.

Nancharaiah, Y. V, Lens, P.N.L., 2015b. Selenium biomineralization for biotechnological applications. Trends Biotechnol. 33, 323–330.

Rath, B.P., Das, S., Mohapatra, P.K. Das, Thatoi, H., 2014. Optimization of extracellular chromate reductase production by Bacillus amyloliquefaciens (CSB 9) isolated from chromite mine environment. Biocatal. Agric. Biotechnol. 3, 35–41.

Rayman, M.P., 2012. Selenium and human health. Lancet 379, 1256–1268.

Sarret, G., Avoscan, L., Carrie, M., Collins, R., Geoffroy, N., Carrot, F., Cove, J., Gouget, B., 2005. Chemical forms of selenium in the metal-resistant bacterium *Ralstonia metallidurans* CH34 exposed to selenite and selenate. Appl. Environ. Microbiol. 71, 2331–2337.

Sasaki, K., Blowes, D.W., Ptacek, C.J., Gould, W.D., 2008. Immobilization of Se(VI) in mine drainage by permeable reactive barriers: column performance. Appl. Geochemistry 23, 1012–1022.

Stams, A.J.M., Grolle, K.C.F., Frijters, C.T.M.J., Van Lier, J.B., 1992. Enrichment of thermophilic propionate-oxidizing bacteria in syntrophy with *Methanobacterium thermoautotrophicum* or *Methanobacterium thermoformicicum*. Appl. Environ. Microbiol. 58, 346–352.

Tan, L.C., Nancharaiah, Y. V, van Hullebusch, E.D., Lens, P.N.L., 2016. Selenium: Environmental significance, pollution, and biological treatment technologies. Biotechnol. Adv. 34, 886–907.

Tinggi, U., 2008. Selenium: Its role as antioxidant in human health. Environ. Health Prev. Med. 13, 102–108.

Tugarova, A. V, Vetchinkina, E.P., Loshchinina, E. a, Burov, A.M., Nikitina, V.E., Kamnev, A. a, 2014. Reduction of selenite by *Azospirillum brasilense* with the formation of selenium nanoparticles. Microb. Ecol. 68, 495–503.

Watts, C. a., Ridley, H., Condie, K.L., Leaver, J.T., Richardson, D.J., Butler, C.S., 2003. Selenate reduction by *Enterobacter cloacae* SLD1a-1 is catalysed by a molybdenum-dependent membrane-bound enzyme that is distinct from the membrane-bound nitrate reductase. FEMS Microbiol. Lett. 228, 273–279.

Wu, L., 2004. Review of 15 years of research on ecotoxicology and remediation of land

147

contaminated by agricultural drainage sediment rich in selenium. Ecotoxicol. Environ. Saf. 57, 257–69.

Yamamura, S., Yamashita, M., Fujimoto, N., Kuroda, M., Kashiwa, M., Sei, K., Fujita, M., Ike, M., 2007. *Bacillus selenatarsenatis* sp. nov., a selenate- and arsenate-reducing bacterium isolated from the effluent drain of a glass-manufacturing plant. Int. J. Syst. Evol. Microbiol. 57, 1060–1064.

Yao, R., Wang, R., Wang, D., Su, J., Zheng, S., Wang, G., 2014. *Paenibacillus selenitireducens* sp. nov., a selenite-reducing bacterium isolated from a selenium mineral soil. Int. J. Syst. Evol. Microbiol. 64, 805–11.

Zhang, Y., Okeke, B.C., Frankenberger, W.T., 2008. Bacterial reduction of selenate to elemental selenium utilizing molasses as a carbon source. Bioresour. Technol. 99, 1267–73.

Zheng, S., Su, J., Wang, L., Yao, R., Wang, D., Deng, Y., Wang, R., Wang, G., Rensing, C., 2014. Selenite reduction by the obligate aerobic bacterium *Comamonas testosteroni* S44 isolated from a metal-contaminated soil. BMC Microbiol. 14, 204.

CHAPTER 7

Selenate bioreduction using methane as electron donor inoculated with marine sediment in a biotrickling filter

Abstract

This study investigated anaerobic bioreduction of selenate to elemental selenium by marine lake sediment from Lake Grevelingen in the presence of methane as a sole electron donor in both batch and continuous studies. Complete bioreduction of 14.3 mg L^{-1} selenate was achieved in batch studies, while up to 143 mg L^{-1} selenate was reduced in the biotrickling filter with a continuous supply of 100% methane gas (in excess) at a reduction rate of 4.12 mg L^{-1} d^{-1}. Red coloured deposits, characteristic of Se(0) particles were observed inside the polyurethane foam packing in the biotrickling filter, which was and were further confirmed using graphite furnace atomic absorption spectrophotometer. The biotrickling filter was operational for 348 days and 100% selenate reduction was observed for the entire time required for complete bioreduction of selenate decreased gradually with each step feed of selenate.

Keywords: Biotrickling filter, selenate, bioreduction, methane, marine lake Grevelingen sediment, selenium wastewater

7.1. Introduction

A range of anthropogenic activities like mining, metallurgy and refining, production of photoelectric devices and semiconductors, power generation and agriculture along with natural weathering and soil leaching contribute towards the enrichment of inorganic selenium (Se) in water streams (Chapman et al., 2009; Tan et al., 2016). Selenium is an essential micronutrient for humans and animals due to its importance in metabolic pathways and immune functions (Zimmerman et al., 2015) and its role in the antioxidant glutathione peroxidase, which protects the cell membrane from damage caused by peroxidation of lipids (Tinggi, 2003). Although Se is an essential trace mineral, an intake at too high concentrations is detrimental leading to Se toxicity (Macfarquhar et al., 2011; Tan et al., 2016). Furthermore, its high rate of bioaccumulation at many trophic levels, even at lower concentrations, makes it a serious human health and environmental concern (Bleiman and Mishael, 2010; Lai et al., 2016; Tan et al., 2016). Chief sources of Se contamination are coal mining and processing, uranium mining, petroleum extraction and refineries and power production (Lemly, 2004; Muscatello et al., 2008). Selenium contamination due to wastewater discharge from mining industry in marine sediment is a recent environmental concern (Ellwood et al., 2016).

Bioreduction of toxic and soluble Se oxyanions to non-toxic and water insoluble Se(0) forms is a promising approach toward tackling the problem of increasing Se contamination in water bodies. However, present studies (Jain et al., 2016, 2015) employ expensive sources of electron donors for selenium bioreduction. It is necessary to explore cheaper and renewable sources of electron donor for selenium bioreduction on larger scale. Methane is one such potential electron donor.

Methane not only occurs naturally in the environment in the form of natural gas, but can also be synthesised renewably (Amon et al., 2007). The International Energy Statistics comprehends that proven reserves stood at 194 trillion cubic metres by the end of 2016 (U.S. E.I.A., 2010). Apart from being a potential energy source, methane is also a potent greenhouse gas (GHG) with a greenhouse warming potential 21 times than that of carbon dioxide (Haynes and Gonzalez, 2014). Anthropogenic sources of methane contribute nearly 20% to the world's GHG warming potential each year when converted to CO_2 (EPA, 2011). Venting and inefficient flaring of natural gas produced as a by-product of petroleum extraction also accounts for 140 trillion cubic meters of GHG released into the atmosphere worldwide in 2011 (Haynes and Gonzalez, 2014). Thus, methane is a cheap and abundantly available carbon source and electron donor for reduction of common sulfur and selenium oxyanions (Bhattarai et al., 2017; Cassarini et al., 2017; Lai et al., 2016).

The sediment biomass used in this study was collected from the marine lake Grevelingen (Scharendijke basin, the Netherlands), from a sulfate-methane transition zone (SMTZ), a region in the sea sediment where the methane rising from below and the sulfate sinking from above form a region suitable for anaerobic methanotrophy (Mcglynn et al., 2015). Recently Bhattarai et al. (2017) studied and confirmed deep sea sediments from SMTZ for possibilities of the reduction of thiosulfate ($S_2O_3^{2-}$), sulfate (SO_4^{2-}) and sulfur (S^0) using ethanol, lactate, acetate and methane as electron donor. Se has a chemical behaviour similar to that of sulfur. Therefore, a selenate reduction similar to sulfate reduction coupled with methane oxidation using the Grevelingen biomass (Bhattarai et al., 2017; Cassarini et al., 2017) was expected in this study.

This research was framed to study anaerobic degradation of selenate in batch bottles using methane and acetate individually as electron donors. Furthermore, a continuous system for bioreduction of selenate with methane as the electron donor and carbon source in a BTF was analysed.

7.2. Materials & methods

7.2.1. Sample collection

The sediment sample was collected from the marine lake Grevelingen, a former Rhine-Meuse estuary on the border of the Dutch provinces of South Holland and Zeeland that has become a lake due to the delta works (Bhattarai et al., 2017; Egger et al., 2016). Sediment coring was done by the NIOZ in Yerseke on the vessel R/V Luctor in June 2015. The location Scharendijke Basin (51°C 44.541'N; 3°C 50.969'E) has a water depth of 45 m and a porosity of 0.85-0.89. For sampling, a UWITEC gravity corer (Mondsee, Austria), equipped with 6 cm inner diameter and 60 cm length core liners, was used. The dark coloured sulphidic sediment (15-20 cm from the top) was homogenized in a nitrogen purged anaerobic chamber (PLAS LABS INCTM) and diluted with an artificial seawater media in a 1:3 ratio and used as marine lake Grevelingen (MLG) sediment seed culture for all experiments.

7.2.2. Medium composition

Artificial seawater media used for the growth of MLG sediment, consisted of NaCl (26 g L^{-1}), $MgCl_2·6H_2O$ (5 g L^{-1}), $CaCl_2·2H_2O$ (1.4 g L^{-1}), NH_4Cl (0.3 g L^{-1}), KH_2PO_4 (0.1 g L^{-1}) and KCl (0.5 g L^{-1}) (Zhang et al., 2010), with 1 mL L^{-1} resazurin solution as redox indicator (0.5 g L^{-1}). The solution with resazurin is colourless at a redox below -110 mV and turns pink at a redox above -51 mV. The following sterile stock solutions described by Widdel and Bak (Balows, 1992) were added to the artificial seawater media: trace element solution (1 mL L^{-1}), 1 M $NaHCO_3$ (30 mL L^{-1}), vitamin mixture (1 mL L^{-1}), thiamine solution (1 mL L^{-1}), vitamin B12 solution (1 mL L^{-1}). The trace element solution comprised of ethylene di amine tetra acetic acid (EDTA) disodium salt (5.2 g L^{-1}), H_3BO_3 (10 mg L^{-1}), $MnCl_2·4H_2O$ (5 mg L^{-1}), $FeSO_4·7H_2O$ (2.1 g L^{-1}), $CoCl_2·6H_2O$ (190 mg L^{-1}), $NiCl_2·6H_2O$ (24 mg L^{-1}), $CuCl_2·2H_2O$ (10 mg L^{-1}), $ZnSO_4·7H_2O$ (144 mg L^{-1}), $Na_2MoO_4·2H_2O$ (36 mg L^{-1}) and the pH of the medium was adjusted to 7.0 ($\pm$ 0.1). The mineral medium was flushed and pressurized with N_2 to remove oxygen.

7.2.3. Batch experiments

Selenate reduction by marine sediment inoculum using different electron donors
Batch studies for selenate reduction were carried out in triplicates by transferring 200 mL of sterile artificial sea water medium in 250 mL airtight serum bottles with thick butyl rubber septa and sealed with aluminium crimps. The serum bottles were made anoxic by purging nitrogen and seeded with 5 mL of the sediment as the source of biomass in an anaerobic chamber.

Different controls containing no selenate, no sediment, heat-killed sediment and no electron donor were each prepared in triplicates. An initial concentration of 14.3 mg L^{-1} of selenate was added to each bottle except for the control without selenate. For batches containing acetate as the electron donor, 250 mg L^{-1} of acetate were added with nitrogen in the head space at 2 bar; while for those containing methane as the electron donor, 100% methane was filled in the headspace at a pressure of 2 bars. The control without electron donor was incubated with 100% nitrogen in the head space at 2 bars. All the batch bottles were incubated in an orbital shaker at 200 rpm at room temperature in the dark for a period of 77 days. The liquid samples were analysed for selenate, total selenium, volatile fatty acids (VFA) and gas samples for methane and carbon dioxide every 2 weeks.

Selenate reduction in the presence of ^{13}C-labelled methane ($^{13}CH_4$)

In order to confirm the occurrence of anaerobic methane oxidation coupled with selenate reduction, the CH_4 oxidation to CO_2 was monitored in the presence of ^{13}C-labelled methane ($^{13}CH_4$) as electron donor. The production of ^{13}C-labelled carbon dioxide ($^{13}CO_2$) was investigated along with selenate removal in batch tests, in the presence of methane as sole electron donor comprising of 5% $^{13}CH_4$ and 95% $^{12}CH_4$. The tests were carried out at atmospheric pressure and different controls as suggested in the previous batch test were performed in triplicates.

7.2.4. Biotrickling filter

A laboratory scale BTF outlined in **Figure 7.1** was constructed to investigate anaerobic selenate bioreduction using methane as an electron donor and consisted of a filter bed encompassed in an airtight glass cylindrical column of 12 cm inner diameter and 50 cm packing height, with a volume of 4.6 L, provided with adequate sampling ports for biomass, gas and liquid sampling. The column was packed with polyurethane foam (98% porosity and a density of 28 kg m^{-3}) cut in cubes of 1-1.5 cm^3, mixed with an equal ratio of plastic pall rings (specific surface area of 188 m^2 m^{-3} and bulk density of 141 kg m^{-3}). The bed was supported by two perforated support plates with 2 mm openings.

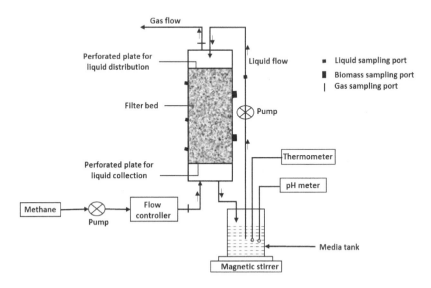

Figure 7.1. Schematic representation of the BTF system for anaerobic reduction of selenate as step feed using methane as electron donor and carbon source

The fluids were circulated in a counter flow with 100% methane in gas phase passed through the column from bottom to top by means of a mass flow controller (SLA 5800 MFC, Brooks instruments) as a continuous feed at a rate of 180 mL h^{-1}. The estimated empty bed residence time for methane was 25.5 h. The artificial seawater medium trickled from the top at a rate of 7.2 L h^{-1}, with a residence time of 1.56 h. The liquid drain was collected into an airtight 3 L glass tank equipped with a mechanical stirrer and a pH probe (pH/mV transmitter - Delta OHM DO9785T) and recirculated to the filter bed through a peristaltic pump (Masterflex L/S Easy-Load II, model 77201-60). The operational parameters are as summarised in **Table 7.1**.

The biotrickling filter was seeded with 500 mL sediment (50% fresh and 50% enriched biomass from the batch studies). 14.3 mg L^{-1} of selenate was added to artificial sea water media in phase one, followed by a 1154 mg L^{-1} of sulfate in phase II to enhance the sulfate reducing activity in the biomass, (Hockin and Gadd, 2003; Y V Nancharaiah and Lens, 2015; Zehr and Oremland, 1987). The concentration of the selenate step-fed to the reactor was gradually increased in each phase and is as summarised in **Table 7.2**.

Table 7.1. Operational parameters of the biotrickling filter

Operational parameters	Conditions
Temperature (°C)	20
pH	7-8
Electron acceptor (mg L^{-1})	Selenate
Carbon source and electron donor	Methane
Growth medium	Artificial seawater
Nutrient loading rate (mL min^{-1})	250
Inlet gas flow rate (mL h^{-1})	180
Volume of BTF bed (L)	4.6
Packing material	Polyurethane foam cubes + plastic pall rings
BTF bed height (cm)	90
Media change	At the beginning of each phase

After 348 days of reactor operation, the Se(0) accumulation in the polyurethane foam cubes was analysed. The foam cubes were washed in Milli-Q water to remove unbound Se and dried in a hot air oven at 70 °C overnight and the weight was recorded. The dry foam cube was cut into tiny pieces of 0.3 – 0.5 mm width, washed in 10 mL 20% HNO_3 (v/v). Washing was repeated serially 10 times. The 10th wash water was analysed for total Se and showed negative for Se, thus indicating that Se was completely extracted from the polyurethane foam cube pieces by serial washing in acidified water. The wash waters from all the 10 washes were mixed together and sonicated at a density of 0.52 W mL^{-1} for 3 minutes to disrupt intact biomass. The liquid sample was then analysed for total Se.

Table 7.2. Phases of operation based on the addition of electron acceptor and selenate removal rates

Operation phase	Days	Selenate (mg L^{-1})	Sulfate (mg L^{-1})	Methane (g m^{-3})	Selenate removal rate (mg L^{-1} d^{-1})
P1	0-24	13.2	0	715.75	1.12
P2	25-56	0	1154.3	715.75	NA
P3	57-97	132	0	715.75	4.12
P4	98-249	13.2	0	715.75	0.79
P5	250-348	39.6	0	715.75	1.45

7.2.5. Analytical methods

The parameters under consideration for this study were selenate, sulfate, CO_2, methane, VFA and total Se. Selenate and sulfate were analysed using ion chromatography (ICS-1000, IC, Dionex with AS-DV sampler) as described by (Cassarini et al., 2017; Dessì et al., 2016). CH_4 and CO_2 in the batch bottles and BTF were measured by gas chromatograph (Scion Instruments 456-GC) fitted with a thermal conductivity detector using He as carrier gas at 1.035 bars as described by Bhattarai et al. (2017). The VFAs were measured by gas chromatograph (Varian 430-GC) with helium as a carrier gas at 1.72 bars. The total Se was analysed in a graphite furnace - atomic absorption spectroscopy (GF-AAS; Solaar AA Spectrometer GF95) as described by (Mal et al., 2016).

Selenate reduction rate and methane utilization were calculated as follows:

The volumetric selenate reduction rate (SeRR) for the BTF at different phases was calculated as follows:

$$SeRR = \frac{[SeO_4^{2-}]_t - [SeO_4^{2-}]_{(t+\Delta t)}}{\Delta t} \qquad \text{(Eq. 7.1)}$$

Percentage methane utilization in the BTF was calculated as follows:

$$\text{Methane utilization (\%)} = \frac{Gas_{in} - Gas_{out}}{Gas_{in}} \times 100 \qquad \text{(Eq. 7.2)}$$

Where $[SeO_4^{2-}]$ = concentration of selenate in mg L^{-1}; t, Δt = time in days; Gas_{in} = Area under the curve for the inlet gas of the BTF in mV min^{-1}; Gas_{out} = Area under the curve for the gas outlet of the BTF in mV min^{-1}.

7.3. Results

7.3.1. Batch studies

Methane as electron donor

Anaerobic selenate removal by MLG sediment was monitored as a function of time with methane as electron donor along with controls with no electron donor, no biomass and killed biomass. Complete bioreduction of 14.3 mg L^{-1} of selenate was observed in about 77 days in the bottles with live biomass, while the controls with no methane and killed biomass, both showed a selenate removal of about 37% (**Figure 7.2**). A thin layer of red coloured particles was observed only in the batch bottles with selenate, methane and live biomass indicating the formation of Se(0) particles (**Figure 7.3**). Control batches without the biomass, and with killed biomass did not show red coloured particles.

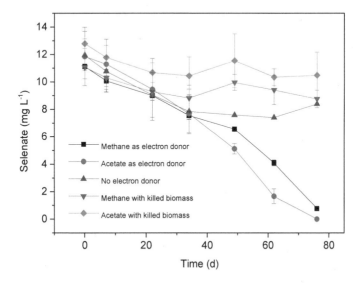

Figure 7.2. Selenate profiles as a function of time with methane and acetate as electron donors, along with controls (killed biomass and the absence of electron donor) showing complete removal of selenate with both methane and acetate as electron donors using BTF biomass sampled on 76 days

Figure 7.3. Thin layer of red coloured Se(0) particle on the media in the serum batch bottles with 14.3 mg L-1 of selenate, methane in the head space at 2 bars and 5 mL MLG sediment in artificial sea water media

Acetate as electron donor

A similar trend of selenate reduction was observed in the serum bottles with acetate as the electron donor. Complete reduction of 14.3 mg L^{-1} of selenate was observed in 80 days in the batches with live biomass, while 23% of selenate removal was observed in the batches with killed biomass and 40.5% in batches with no electron donor without externally supplied electron donor (**Figure 7.4**). For the selenate reduction studies, an initial concentration of 250 mg L^{-1} of acetate was used and acetate was removed with the course of time, indicating that acetate was indeed used as electron donor for selenate reduction. This is further supported by the fact that acetate was not being consumed in the batches with killed biomass.

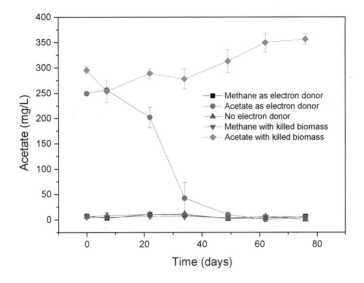

Figure 7.4. Acetate consumption profile with live and killed biomass showing complete degradation of acetate in 50 days in batches with live biomass and no degradation in batches with autoclaved biomass

[13]C-labelled methane as electron donor

In contrast to the previous experiment, no selenate removal from each batch bottles in the presence of $^{13}CH_4$ as sole source of electron donor was observed (**Figure 7.5a**). On day 7, a steep decrease in concentration of selenate in each bottle was observed. This could be attributed to the absorption of selenate on the sediments or due to sampling/analytical error. No significant change was observed in selenate in medium and Se in sediment observed after 35 days. On day 45 and 55, slight decrease in selenate concentration in the medium was observed in the experimental set-ups with ^{13}C and ^{12}C methane in the headspace but also in control without methane in the headspace (headspace with N_2). This removal of selenate from the medium could be due to oxidation of the organic carbon content in the sediment and not because of oxidation of ^{13}C-labelled methane. Similarly, the changes in the ^{13}C-labelled carbon dioxide (**Figure 7.5b**) was similar to that observed in the test and control bottles, suggesting that the methane was not utilised for the selenate reduction in the batch experiments.

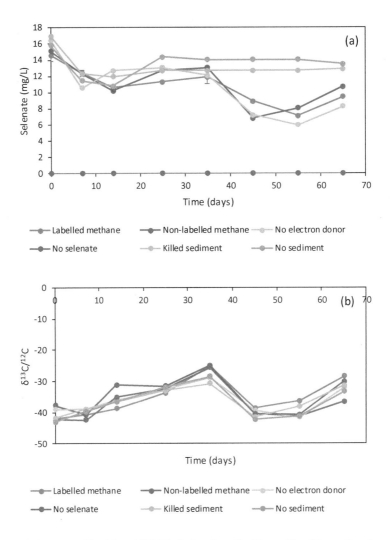

Figure 7.5. Selenate profiles (a) and ¹³C-labelled carbon dioxide profiles (b) as a function of time with ¹³C-labelled methane and ¹²C methane as electron donors, along with controls (killed biomass and the absence of electron donor)

7.3.2. Continuous studies in biotrickling filter

Selenate and sulfate bioreduction

Selenate bioreduction was evaluated in the biotrickling filter as a function of time by dividing the total study period of 348 days into 5 phases (**Table 7.2**), based on the dosage of selenate and sulfate. 100% removal of both selenate and sulfate was achieved in the entire 348 days study. A removal rate ranging from 0.312-4.123 mg L^{-1} d^{-1} of selenate and 32.48 mg L^{-1} d^{-1} of sulfate were observed (**Figure 7.6**). The average removal rates of selenate for each phases is summarised in **Table 7.2**. In phase P4 and P5, the average removal rates for each step feed of selenate increased with each run. Red colour precipitates, characteristic of elemental Se(0) (Lenz et al., 2008), were observed on the sponges at the end of BTF run (**Figure 7.6**). The total Se concentration in the sponges was measured.

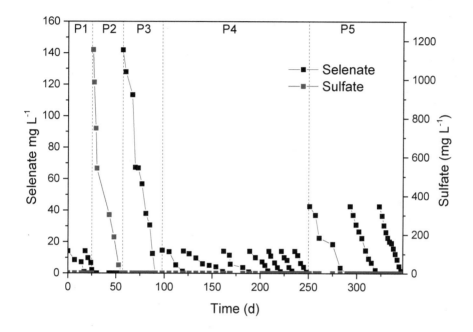

Figure 7.6. Time profile of selenate and sulfate in the biotrickling filter phased out based on the initial step feed concentration showing complete removal of sulfate and selenate. The vertical lines represent different phases in the BTF operation; P1: days 0-24, P2: days 25-56, P3: days 57-97, P4: days 98-249, P5: days 250-348. Artificial sea water media was changed at the beginning of each phase.

Acetate and propionate production and methane utilization

From the beginning of phase 4 at 98 days of operation, the VFA profiles were monitored to account for the high selenate removal rate of 4.12 mg L^{-1} d^{-1} in phase P3. Concentrations up to 60 mg L^{-1} of propionate were observed which eventually decreased to non-detectable by 175 days of operation. In the case of acetate, 580 mg L^{-1} was observed at the beginning of phase P4 and decreased to 185 mg L^{-1} after 245 days of operation and eventually to non-detectable limits after 280 days of operation (**Figure 7.7a, 7.7b**). The batch tests with acetate as electron donor showed that the biomass was capable of using short chain fatty acids as a carbon source (**Figure 7.4**). The inlet and outlet methane concentration profiles were monitored from phase P4 starting from 98 days of operation. The measurements were not made in P1 to P3 because of technical limitations. Average methane utilization in the reactor was 8.52% with a maximum utilization of 16-6%, at 263 days of operation in phase P5 (**Figure 7.7c**).

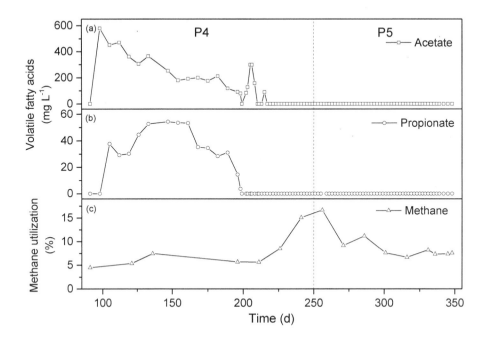

Figure 7.7. Time profile of (a) acetate and (b) propionate concentrations along with (c) Methane utilization (%) in the BTF in phases P4 and P5 The vertical line separates phase P4: days 98-249 and P5: days 250-348.

Figure 7.8. Comparison of polyurethane foam cubes before and after BTF operation: (a) Fresh polyurethane foam cubes before the operation of BTF. (b) Polyurethane foam cubes from the reactor after 348 days of operation showing adsorbed Se(0) in red colour.

7.4. Discussion

7.4.1. Bioreduction of selenate coupled with anaerobic oxidation of methane

This study concludes for the first time that methane could be used as electron donor for complete bioreduction of selenate in a BTF for a long term operation of 348 days. During the entire 348 days, no operational problems such as biomass over growth, filter clogging was encountered. Selenium oxyanion reduction by chemical and biological processes have been well documented in various studies (Howarth et al., 2015; Yarlagadda V Nancharaiah and Lens, 2015; Subedi et al., 2017; Tan et al., 2016; Tang et al., 2014) that include selenate co-reduction with other electron acceptors like sulfate (Zehr and Oremland, 1987) and ammonia (Mal et al., 2017) using different electron donors (Losi et al., 1997; Viamajala et al., 2006). Chun-Yu Lai et al. in their work recorded co-bioreduction of selenate and nitrate using methane as an electron acceptor in a membrane biofilm reactor operational for 140 days (Lai et al., 2016). A stable bioreduction of concentrations up to 143 mg L^{-1} as a shock load and repeated cycles of 14.3 mg L^{-1} and 42.9 mg L^{-1} were efficiently reduced without the system crashing and long term operation for 348 days (**Figure 7.6**).

Few studies on reduction of selenium oxyanions explore the possibility of using gas phase electron donor, especially an inexpensive source like methane (Cassarini et al., 2017) and mostly focus on liquid phase substrates (Lai et al., 2016; Mal et al., 2017; Yarlagadda V Nancharaiah and Lens, 2015), while in this work the feasibility of methane as an electron donor was investigated. In recent years, methane has become a very abundantly available resource due to the discovery of new geographical sources and world wide application of anaerobic digestion process (Amon et al., 2007; Leahy et al., 2013). Batch experiments showed that methane was used as the electron donor for complete bioreduction of 14.3 mg L^{-1} selenate in 77 days (**Figure 7.2**). In contrast, the controls without methane showed only 39.3% reduction. To further confirm the role of the biomass, the controls with no sediment showed 36.6% of selenate bioreduction in the same period of time. This could be attributed to the change in volume in the batches after each sampling. This may be avoided by using sacrificial batches that can be discarded after each sampling.

In order to confirm the process of anaerobic methane oxidation coupled with selenate reduction, ^{13}C-labelled methane was added as sole electron donor in the batch experiments. However, no selenate removal was observed during the process (**Figure 7.5a**). While the previous experiments were performed under 2 bar pressure of ^{12}CH$_4$, this study with ^{13}C-labelled methane were performed at atmospheric pressure. Dissolution of methane is significantly affected by the pressure conditions in the batch tests, which could be the possible explanation for no selenate reduction in the batch tests. However, further study with ^{13}C-labelled methane was added as sole electron donor under 2 bar pressure should be carried out to understand the effect of pressure on selenate reduction.

Selenate being a component of wastewaters in petroleum refinery waste waters (Leahy et al., 2013; Tan et al., 2016), a process for co-removal, where methane oxidation could complement a process like selenate reduction would be effective. Though the reactor was fed with 100% methane at a rate of 180 mL/h (**Table 7.1**), the bioavailable fraction would be low considering the fact that the filter bed is completely saturated with water and the methane is available for the organisms only after dissolving in the liquid phase surrounding the biofilm, and the solubility of methane in seawater is 0.055 mg/m^3 (Duan and Mao, 2006; Serra et al., 2006), making the bioavailability of methane a limiting factor.

7.4.2. Acetate and propionate production in BTF

Along with methane, a set of batches with acetate as electron donor were carried out. Acetate being a well-studied candidate as a substrate for bioreduction of various chalcogen oxyanions (Bhattarai et al., 2017). The batch results confirm that the marine sediment are capable of selenate reduction (**Figure 7.2**), coupled with acetate utilization (**Figure 7.4**). To account for high selenate bioreduction rates in phase P4, the reactor was monitored for VFA production and acetate and propionate at concentrations of 580 mg L^{-1} and 185 mg L^{-1} were observed. This suggests that methane was oxidised to acetate and propionate. Chun-Yu Lai et al. also observed acetate production in their work on selenate reduction using methane as electron donor (Lai et al., 2016). Eventually, the concentrations reduced to non-detectable amounts in the course of the long operational period. This can be supported by the use of acetate by the biomass for selenate reduction in the batch studies (**Figure 7.4**).

The topic of anaerobic oxidation of methane to acetate and other hydrocarbons is widely discussed as an economically significant process and is becoming attractive for the potential it holds towards alternate fuel technology (Chowdhury and Maranas, 2015; Fei et al., 2014; Jin and Yeol, 2016; Mueller et al., 2015; Nazem-Bokaee et al., 2016; Yang et al., 2014). The Thermofischer's process for thermochemical conversion of methane to alcohols and higher hydrocarbons (Lunsford, 2000; Miller and Sorrell, 2014) is established and similar conversions (biological) are being explored. Soo et al. in their studies report the occurrence of reverse methanogenesis and conversion of methane to acetate by expression of enzyme methyl-coenzyme M reductase (Mcr) from ANME-1, cloned in a methanogenic host (Soo et al., 2016). However, the speculation of acetate and propionate production in the reactor needs further confirmation with dedicated analysis studies using labelled methane studies and NanoSIMS.

7.4.3. Practical implications

Slower growth rate and doubling time of the biomass from Lake Grevelingen lead to long term operation of the batches irrespective of the electron donor provided. Other studies (Espinosa-Ortiz et al., 2015; Jain et al., 2016, 2015; Liang et al., 2015; Subedi et al., 2017; Tang et al., 2014) show removal of similar concentrations of selenium oxyanions within the period of 1-2 weeks using anaerobic granular sludge. However, the advantage of using MLG sediment is that the biomass is capable of utilizing inexpensive and readily available methane as electron donor. Long term and continuous operation of the process using methane as the electron donor might help to increase the reduction rate of the selenium oxyanions and also reduce the operation cost. Methane and selenate being common pollutants from the petrochemical refineries, a process for utilization of methane for bioreduction of selenate, as illustrated in this study, is an interesting option. During this study, it was observed that the sponges in the filter bed gradually turned red and brighter in the course of the operation. Further GF-AAS analysis of randomly collected sponges confirmed that accumulation of $3.8 - 8.02$ mg Se g^{-1} of sponge (dry weight) which could represent an added value for the process through resource recovery.

In the order of abundance, Se ranks 69th in the earth's crust. Se has many industrial applications like semiconductors and photoelectric cells, metallurgy and glass industries (Macaskie et al., 2010; Tan et al., 2016) and healthcare applications. Increasing the industrial demand of Se may be approached with its recovery from wastewater streams. In this study, an inexpensive and continuous bioprocess for selenium removal and recovery has been explored and must be further pursued on a larger scale. Although, there are a vast number of studies on reduction of Se oxyanions reduction, there is seldom any focus on accumulation and recovery of the reduced form. In this study, Se(0) particles were accumulated in the filter bed and could be recovered. The serial washing of the polyurethane foam cubes in acidified water and analysis for total selenium revealed the accumulation of $3.8 - 8.02$ mg Se g^{-1} sponge (dry weight) in the filter bed. Thus, avoiding the carryover of Se further into the environment where oxidation to reduced Se forms is very likely.

7.5. Conclusion

Complete bioreduction of selenate using methane as an electron donor was demonstrated in this study using biomass from a deep sea sediment collected from a sulfur-methane transition zone. Methane is a good candidate for bioreduction of selenate to elemental Se using biomass from the marine sediment. The rate of bioreduction consistently increased with each cycle of operation and Se(0), which is the end product of the bioreduction accumulated in the filter bed A microbial community analysis for the enriched biomass would further broaden the understanding of the biochemical pathways, acetate and propionate production and utilization.

References

Amon, T., Amon, B., Kryvoruchko, V., Machmüller, A., Hopfner-Sixt, K., Bodiroza, V., Hrbek, R., Friedel, J., Pötsch, E., Wagentristl, H., Schreiner, M., Zollitsch, W., 2007. Methane production through anaerobic digestion of various energy crops grown in sustainable crop rotations. Bioresour. Technol. 98, 3204–3212.

Balows, A., 1992. The Prokaryotes: a handbook on the biology of bacteria: ecophysiology, isolation, identification, applications. Springer New York, New York, USA.

Bhattarai, S., Cassarini, C., Naangmenyele, Z., Rene, E.R., 2017. Microbial sulfate-reducing activities in anoxic sediment from Marine Lake Grevelingen : screening of electron donors and acceptors. Limnology. doi:10.1007/s10201-017-0516-0

Bleiman, N., Mishael, Y.G., 2010. Selenium removal from drinking water by adsorption to chitosan-clay composites and oxides: Batch and columns tests. J. Hazard. Mater. 183, 590–595.

Cassarini, C., Rene, E.R., Bhattarai, S., Esposito, G., Lens, P.N.L., 2017. Anaerobic oxidation of methane coupled to thiosulfate reduction in a biotrickling filter. Bioresour. Technol. 240, 214–222.

Chapman, P.M., Adams, W.J., Brooks, M.L., Delos, C., Luoma, S.N., Maher, W.A., Olhendorf, H.M., Presser, T.S., Shaw, D.P., 2009. Ecological assessment of selenium in the aquatic environment: Summary of a SETAC Pellston workshop. Pensacola FL (USA).

Chowdhury, A., Maranas, C.D., 2015. Designing overall stoichiometric conversions and intervening metabolic reactions. Sci. Rep. 5, 16009.

Tan, L.C., Nancharaiah, Y. V, van Hullebusch, E.D., Lens, P.N.L., 2016. Selenium: Environmental significance, pollution, and biological treatment technologies. Biotechnol. Adv. 34, 886–907.

Dessì, P., Jain, R., Singh, S., Seder-Colomina, M., van Hullebusch, E.D., Rene, E.R.,

Ahammad, S.Z., Carucci, A., Lens, P.N.L., 2016. Effect of temperature on selenium removal from wastewater by UASB reactors. Water Res. 94, 146–154.

Dorer, C., Vogt, C., Neu, T.R., Stryhanyuk, H., Richnow, H.-H. 2016. Characterization of toluene and ethylbenzene biodegradation under nitrate-, iron (III)-and manganese (IV)-reducing conditions by compound-specific isotope analysis. *Environ. Pollut.*, 211, 271-281.

Duan, Z., Mao, S., 2006. A thermodynamic model for calculating methane solubility, density and gas phase composition of methane-bearing aqueous fluids from 273 to 523 K and from 1 to 2000 bar. Geochim. Cosmochim. Acta 70, 3369–3386.

Egger, M., Lenstra, W., Jong, D., Meysman, F.J.R., Sapart, C.J., 2016. Rapid sediment accumulation results in high methane effluxes from coastal sediments. PLoS One 1–22. doi:10.1594/PANGAEA.863726

Ellwood, M.J., Schneider, L., Potts, J., Batley, G.E., Floyd, J., Maher, W.A., 2016. Volatile selenium fluxes from selenium-contaminated sediments in an Australian coastal lake. Environ. Chem. 13, 68–75.

EPA, 2011. Global Anthropogenic Non-CO 2 Greenhouse Gas Emissions : 1990 - 2030, Office of Atmospheric Programs Climate Change Division U.S. Environmental Protection Agency. NW Washington, DC.

Espinosa-Ortiz, E., Rene, E.R., van Hullebusch, E.D., Lens, P.N.L., 2015. Removal of selenite From wastewater in a *Phanerochaete chrysosporium* pellet based fungal bioreactor. Int. Biodeterior. Biodegradation 102, 361–369.

Fei, Q., Guarnieri, M.T., Tao, L., Laurens, L.M.L., Dowe, N., Pienkos, P.T., 2014. Bioconversion of natural gas to liquid fuel: Opportunities and challenges. Biotechnol. Adv. 32, 596–614.

Haynes, C. a, Gonzalez, R., 2014. Rethinking biological activation of methane and conversion to liquid fuels. Nat. Chem. Biol. 10, 331–9.

Hockin, S.L., Gadd, G.M., 2003. Linked redox precipitation of sulfur and selenium under anaerobic conditions by sulfate-reducing bacterial biofilms. Appl. Enivironmental Microbiol. 69, 7063–72.

Howarth, A.J., Katz, M.J., Wang, T.C., Platero-prats, A.E., Chapman, K.W., Hupp, J.T., Farha, O.K., 2015. High efficiency adsorption and removal of selenate and selenite from water using metal − organic frameworks. J. Am. Chem. Soc. 137(23), 7488-94.

Jain, R., Matassa, S., Singh, S., van Hullebusch, E.D., Esposito, G., Lens, P.N.L., 2016. Reduction of selenite to elemental selenium nanoparticles by activated sludge. Environ.

Sci. Pollut. Res. 23, 1193–1202.

Jain, R., Seder-Colomina, M., Jordan, N., Dessi, P., Cosmidis, J., van Hullebusch, E.D., Weiss, S., Farges, F., Lens, P.N.L., 2015. Entrapped elemental selenium nanoparticles affect physicochemical properties of selenium fed activated sludge. J. Hazard. Mater. 295, 193–200.

Jin, T., Yeol, E., 2016. Metabolic versatility of microbial methane oxidation for biocatalytic methane conversion. J. Ind. Eng. Chem. 35, 8–13.

Lai, C.-Y., Wen, L.-L., Shi, L.-D., Zhao, K.-K., Wang, Y.-Q., Yang, X., Rittmann, B.E., Zhou, C., Tang, Y., Zheng, P., Zhao, H.-P., 2016. Selenate and nitrate bioreductions using methane as the electron donor in a membrane biofilm reactor. Environ. Sci. Technol. 50, 10179–86.

Leahy, M., Barden, J.L., Murphy, B.T., Slater-thompson, N., Peterson, D., 2013. International Energy Outlook 2013.

Lemly, A.D., 2004. Aquatic selenium pollution is a global environmental safety issue. Ecotoxicol. Environ. Saf. 59, 44–56.

Lenz, M., Hullebusch, E.D. Van, Hommes, G., Corvini, P.F.X., Lens, P.N.L., 2008. Selenate removal in methanogenic and sulfate-reducing upflow anaerobic sludge bed reactors. Water Res. 42, 2184–94.

Liang, L., Guan, X., Huang, Y., Ma, J., Sun, X., Qiao, J., 2015. Efficient selenate removal by zero-valent iron in the presence of weak magnetic field 156, 1064–1072.

Losi, M.E., Frankenberger, W.T., Valley, S.J., 1997. Reduction of selenium oxyanions by *Enterobacter cloacae* SLD1a-1 : Isolation and growth of the bacterium and its expulsion of selenium particles 63, 3079–84.

Lunsford, J.H., 2000. Catalytic conversion of methane to more useful chemicals and fuels: A challenge for the 21st century. Catal. Today 63, 165–74.

Macaskie, L.E., Mikheenko, I.P., Yong, P., Deplanche, K., Murray, A.J., Paterson-Beedle, M., Coker, V.S., Pearce, C.I., Cutting, R., Pattrick, R.A.D., Vaughan, D., van der Laan, G., Lloyd, J.R., 2010. Today's wastes, tomorrow's materials for environmental protection. Hydrometallurgy 104, 483–487.

Macfarquhar MJK, Broussard DL, Burk RF, Dunn JR, Green AL (2011) Acute Selenium Toxicity Associated With a Dietary Supplement. Arch. Int. Med. 170(3), 256–261.

Mal, J., Nancharaiah, Y. V., van Hullebusch, E.D., Lens, P.N.L., 2016. Effect of heavy metal co-contaminants on selenite bioreduction by anaerobic granular sludge. Bioresour. Technol. 206, 1–8.

Mal, J., Nancharaiah, Y. V, Hullebusch, E.D. Van, Lens, P.N.L., 2017. Biological removal of selenate and ammonium by activated sludge in a sequencing batch reactor. Bioresour. Technol. 229, 11–19.

Mcglynn, S.E., Chadwick, G.L., Kempes, C.P., Orphan, V.J., 2015. Single cell activity reveals direct electron transfer in methanotrophic consortia. Nature 526, 531–535.

Miller, R.G., Sorrell, S.R., 2014. The future of oil supply. Phil. Trans. R. Soc. A. 372, 20130179.

Mueller, T.J., Grisewood, M.J., Nazem-Bokaee, H., Gopalakrishnan, S., Ferry, J.G., Wood, T.K., Maranas, C.D., 2015. Methane oxidation by anaerobic archaea for conversion to liquid fuels. J. Ind. Microbiol. Biotechnol. 42, 391–401.

Muscatello, J.R., Belknap, A.M., Janz, D.M., 2008. Accumulation of selenium in aquatic systems downstream of a uranium mining operation in northern Saskatchewan , Canada. Environ. Pollut. 156, 387–393.

Nancharaiah, Y. V, Lens, P.N.L., 2015. Ecology and biotechnology of selenium-respiring bacteria. Microbiol. Mol. Biol. Rev. 79, 61–80.

Nancharaiah, Y. V, Lens, P.N.L., 2015. Selenium biomineralization for biotechnological applications. Trends Biotechnol. 33, 323–330.

Nazem-Bokaee, H., Gopalakrishnan, S., Ferry, J.G., Wood, T.K., Maranas, C.D., 2016. Assessing methanotrophy and carbon fixation for biofuel production by Methanosarcina acetivorans. Microb. Cell Fact. 15, 10.

Serra, M.C.C., Pessoa, F.L.P., Palavra, A.M.F., 2006. Solubility of methane in water and in a medium for the cultivation of methanotrophs bacteria. J. Chem. Thermodyn. 38, 1629–1633.

Soo, V.W.C., McAnulty, M.J., Tripathi, A., Zhu, F., Zhang, L., Hatzakis, E., Smith, P.B., Agrawal, S., Nazem-Bokaee, H., Gopalakrishnan, S., Salis, H.M., Ferry, J.G., Maranas, C.D., Patterson, A.D., Wood, T.K., 2016. Reversing methanogenesis to capture methane for liquid biofuel precursors. Microb. Cell Fact. 15, 11.

Subedi, G., Taylor, J., Hatam, I., Baldwin, S.A., 2017. Chemosphere Simultaneous selenate reduction and denitri fi cation by a consortium of enriched mine site bacteria. Chemosphere 183, 536–545.

Zehr JP, Oremland RS. 1987. Reduction of Selenate to Selenide by Sulfate-Respiring Bacteria: Experiments with Cell Suspensions and Estuarine Sediments. Appl. Environ. Microbiol. 53(6), 1365-1369.

Tan, L.C., Nancharaiah, Y. V, van Hullebusch, E.D., Lens, P.N.L., 2016. Selenium:

Environmental significance, pollution, and biological treatment technologies. Biotechnol. Adv. 34, 886–907.

Tang, C., Huang, Y.H., Zeng, H., Zhang, Z., 2014. Reductive removal of selenate by zero-valent iron : The roles of aqueous Fe(2+) and corrosion products , and selenate removal mechanisms. Water Res. 67, 166–174.

Tinggi, U., 2003. Essentiality and toxicity of selenium and its status in Australia : a review. Toxicol. Lett. 137, 103–110.

U.S. E.I.A., 2010. International Energy Statistics.

Viamajala, S., Bereded-Samuel, Y., Apel, W.A., Petersen, J.N., 2006. Selenite reduction by a denitrifying culture: Batch- and packed-bed reactor studies. Appl. Microbiol. Biotechnol. 71, 953–962.

Yang, L., Ge, X., Wan, C., Yu, F., Li, Y., 2014. Progress and perspectives in converting biogas to transportation fuels. Renew. Sustain. Energy Rev. 40, 1133–52.

Zhang, Y., Henriet, J.P., Bursens, J., Boon, N., 2010. Stimulation of in vitro anaerobic oxidation of methane rate in a continuous high-pressure bioreactor. Bioresour. Technol. 101, 3132–38.

Zehr JP, Oremland RS. 1987. Reduction of Selenate to Selenide by Sulfate-Respiring Bacteria: Experiments with Cell Suspensions and Estuarine Sediments. Appl. Environ. Microbiol. 53(6), 1365-69.

Zimmerman, M.T., Bayse, C.A., Ramoutar, R.R., Brumaghim, J.L., 2015. Sulfur and selenium antioxidants : Challenging radical scavenging mechanisms and developing structure – activity relationships based on metal binding. J. Inorg. Biochem. 145, 30–40.

CHAPTER 8

Formation of Se(0), Te(0) and Se(0)-Te(0) nanostructures during simultaneous bioreduction of selenite and tellurite in upflow anaerobic sludge blanket reactor

This chapter has been modified and published as:

Wadgaonkar S.L., Mal J., Nancharaiah Y.V., Maheshwari N.O., Esposito G., Lens P.N.L. 2018. Formation of Se(0), Te(0) and Se(0)-Te(0) nanostructures during simultaneous bioreduction of selenite and tellurite in a UASB reactor. Appl. Microbiol. Biotechnol. 102(6): 2899-2911. DOI: 10.1007/s00253-018-8781-3.

Abstract

Simultaneous removal of selenite and tellurite from synthetic wastewater was achieved through microbial reduction in a lab-scale upflow anaerobic sludge blanket reactor operated with 12 h hydraulic retention time at 30 °C and pH 7 for 120 days. Lactate was supplied as electron donor at an organic loading rate of 528 or 880 mg COD L^{-1} day^{-1}. The reactor was initially fed with a synthetic influent containing 0.05 mM selenite and tellurite each (phase I, day 1-60) and subsequently with 0.1 mM selenite and tellurite each (phase II, day 61-120). At the end of phase I, selenite and tellurite removal efficiencies were 93 and 96%, respectively. The removal percentage dropped to 87% and 81% for selenite and tellurite, respectively, at the beginning of phase II because of the increased influent concentrations. The removal efficiencies of both selenite and tellurite were gradually restored within 20 days and stabilized at ≥97% towards the end of the experiment. Powder X-ray diffraction and Raman spectroscopy confirmed the formation of biogenic Se(0), Te(0) and Se(0)-Te(0) nanostructures. Scanning transmission electron microscopy coupled with energy-dispersive X-ray spectroscopy showed aggregates comprising of Se(0), Te(0) and Se-Te nanostructures embedded in a layer of extracellular polymeric substances (EPS). Infrared spectroscopy confirmed the presence of chemical signatures of the EPS which capped the nanoparticle aggregates that had been formed and immobilized in the granular sludge. This study suggests a model for technologies for remediation of effluents containing Se and Te oxyanions coupled with biorecovery of bimetal(loid) nanostructures.

Keywords: Biogenic selenium; biogenic nanoparticles; bioremediation; selenite reduction; tellurite reduction; wastewater treatment.

8.1. Introduction

Selenium (Se) and tellurium (Te) belong to the chalcogen group, i.e. group 16 of the periodic table. These elements are found in nature most commonly associated with copper- and sulfur-bearing ores (George 2003; Spinks et al. 2016). Both Se and Te exist in nature in four major oxidation states: +VI, +IV, 0 and –II. The oxyanions (+VI and +IV) of both Se and Te are soluble and mobile, while the elemental forms (Se(0) and Te(0)) are insoluble in water (Chasteen and Bentley 2003; Turner et al. 2012). Among the naturally occurring minerals, selenium and tellurium are relatively low in abundance in the Earth's crust and are ranked 69th and 75th, respectively (Dhillon and Dhillon 2003; Yoon et al. 1990). In order to ensure sustainability of these elements, development of new technologies for the recovery of both Se and Te from waste streams are essential.

Use of microorganisms in developing biotechnological processes for removal and recovery of critical elements (e.g. Te, Co and Pd) is becoming more common because of high efficiency, cost effectiveness and environmentally friendlier approach (Mal et al. 2016b; Nancharaiah et al. 2016). Removal and recovery of these critical elements from wastes and contaminated soil-water environmental settings has a dual benefit: 1) bioremediation of the contaminated soil-water environment or treatment of effluents, and 2) recovered elements can be considered for re-use to partially meet the constantly growing industrial demand (Jorgenson 2002). A large number of studies is available on the microbial reduction of selenium oxyanions to insoluble Se(0) (Hunter 2014; Li et al. 2014; Staicu et al. 2015). In recent years, microbial reduction of tellurium oxyanions and recovery of Te(0) has also become of interest (Baesman et al. 2007; Espinosa-Ortiz et al. 2017; Kagami et al. 2012; Rajwade and Paknikar 2003). However, few reports have appeared that deal with simultaneous biological removal of Se and Te oxyanions from wastewaters that contain both Se and Te contaminants (Bajaj and Winter 2014; Espinosa-Ortiz et al. 2017).

Anthropogenic activities such as mining and refining of certain ores can lead to contamination of soil-water environments with both Se and Te. The eighty-year-old nickel refinery in Swansea Valley (UK) is an example where nearby soil and water are polluted with both Se and Te (Perkins 2011). In the topsoil surrounding this nickel refinery, Se and Te concentrations as high as 200 mg kg^{-1} and 11 mg kg^{-1}, respectively, were recorded. Se and Te are recovered as contaminants of copper ore by the copper mining industry (Jorgenson 2002). Around 41,000 metric tons of copper anode slime is generated annually by the mining industries (Jorgenson 2002). But 17% of this anode slime is discarded into the wastewater streams because refineries do not possess the equipment or the financial means for processing it. Anode slime produced in the primary metal refineries contains about 10-40% selenium and up to 5% tellurium (Jorgenson 2002).

If Se and Te are present as co-contaminants, the presence of one oxyanion may influence microbial reduction of the other oxyanion and *vice versa*. Bajaj and Winter (2014) recently reported that the tellurite reduction rate by aerobically growing *Duganella violacienigra* increased 13 fold in the presence of selenite. Simultaneous reduction of both Se and Te oxyanions may present opportunities for biogenesis of Se-Te nanostructures (Bajaj and Winter 2014; Espinosa-Ortiz et al. 2017). Compared to pure Se and Te nanomaterials, nanodevices based on Se-Te alloys have superior properties such as electrical resistance and magnetoresistance (Fu et al. 2015; Jian and Yadong 2003; Mayers et al. 2001).

In previous studies, formation of Se(0) and Te(0) nanostructures were noted when axenic cultures of bacteria (Bajaj and Winter 2014) or fungi (Espinosa-Ortiz et al. 2017) were incubated with both Se(IV) and Te(IV) oxyanions. Possibilities for formation of Se-Te nanostructures were also discussed. However, no reports have appeared so far on studies of simultaneous selenite and tellurite reduction by mixed microbial communities (i.e., biofilms or granular sludge) under anaerobic conditions that are suitable for process development in bioreactors. Practical application of large-scale recovery of Se/Te would require a continuous and cost-effective technology. Upflow anaerobic sludge blanket (UASB) reactors are most commonly employed for high-rate anaerobic wastewater treatment (Dessì et al. 2016; Lenz et al. 2008). Advantages of utilizing UASB reactors include energy recovery, reduced maintenance and operation costs and easy recovery of the biomass associated nanoparticles for future reuse (Mal et al. 2017b; Pat-Espadas et al. 2016; Ramos-Ruiz et al. 2017).

The primary objective of this study was to investigate the simultaneous reduction of selenite and tellurite by anaerobic sludge in a lab-scale UASB reactor and the characterization of the biogenic Se(0), Te(0), and Se-Te nanostructures. To our knowledge, this is the first report on simultaneous reduction of selenite and tellurite by mixed microbial consortia in a UASB bioreactor. Constant removal of selenite and tellurite along with retention of biogenic Se and Te nanostructures in the bioreactor was investigated during a 120-day long operation. Biogenic Se and Te structures associated with the granular sludge were characterized using transmission electron microscopy (TEM), scanning transmission electron microscopy coupled with energy-dispersive X-ray spectroscopy (STEM-EDS), powder X-ray diffraction (P-XRD), Raman spectroscopy and attenuated total reflection infra-red (ATR-IR) spectroscopy. In addition, recovery of the biogenic Se, Te and Se-Te nanostructures from selenite and tellurite reducing granular sludge was investigated.

8.2. Materials and methods

8.2.1. Source of biomass

Anaerobic granular sludge collected from a full-scale UASB reactor, kindly provided by Biothane Systems International B.V. (Delft, the Netherlands), in which brewery wastewater was being treated (Gonzalez-Gil et al. 2001; Mal et al. 2017b), was used as inoculum for both the batch experiments and the lab-scale UASB reactor. The microbial community composition of methanogenic granular sludge was δ-*Proteobacteria* (33%) as the most abundant bacterial class, followed by *Bacteroidia* (15%), *Clostridia* (10%) and *Anaerolinea* (10%) (Gonzalez-Gil et al. 2016).

8.2.2. Synthetic wastewater

The composition of the synthetic wastewater used in this study was as follows (in g L^{-1}): $Na_2HPO_4.2H_2O$ (0.053), KH_2PO_4 (0.041), NH_4Cl (0.300), $CaCl_2.2H_2O$ (0.010), $MgCl_2.6H_2O$ (0.010) and $NaHCO_3$ (0.040).

During phase I, lactate was added to a concentration of 3 mM (264 mg COD L^{-1}) and during phase II, it was added to a concentration of 5 mM (440 mg COD L^{-1}). These lactate concentrations corresponded to organic loading rates of 528 and 880 mg COD L^{-1} day^{-1}, respectively. The lactate served as carbon source and electron donor. A volume of 0.1 mL of acid and of alkaline trace element solutions were added to 1 L of synthetic wastewater medium as described by Stams et al. (1992). The resultant pH of the synthetic wastewater was 7.0 - 7.1.

8.2.3. Batch incubations with various selenium and tellurium concentrations

Batch experiments were performed with the synthetic wastewater supplemented with either 1 mM of tellurite or 1 mM selenite. Duplicate serum bottles containing medium with either oxyanion and 10 mM lactate as electron donor were inoculated with either 2 or 3 g of moist anaerobic granular sludge (dry weight corresponding to 13.4% (w/w) wet weight; 2.5 g L^{-1} volatile solids (VS); 4.6 g L^{-1} total solids).

Another set of batch experiments was performed to study the effect of selenite on the reduction of tellurite. Duplicate bottles with wastewater medium containing 10 mM lactate and 1 mM tellurite and 1 mM selenite each were inoculated with 2 or 3 g wet anaerobic granular sludge. In all the foregoing experiments, the serum bottles were fitted with rubber stoppers and sealed with aluminium crimps. The serum bottles were purged with ultra high pure nitrogen gas for 5 min and incubated at 30 °C on an orbital shaker at 180 rpm. Liquid samples were withdrawn anaerobically at regular time intervals using a syringe for monitoring selenite, tellurite and lactate concentrations.

8.2.4. UASB reactor operation

A UASB reactor (**Figure 8.1**), made of a transparent polyvinyl cylinder, had a height and diameter of 70 cm and 5.7 cm, respectively, and a working volume of 1.2 L. The reactor was operated at an upflow velocity of 0.48 m day^{-1}, organic loading rate (OLR) based on chemical oxygen demand (COD) of 528 or 880 mg L^{-1} day^{-1} and a hydraulic retention time (HRT) of 12 h. The reactor was operated at 30.0 (± 0.5) °C and pH 7.

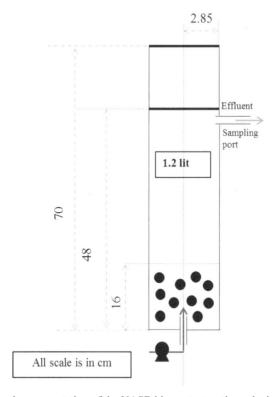

Figure 8.1 Schematic representation of the UASB bioreactor treating selenite and tellurite containing wastewater.

The UASB reactor was operated in 2 phases. In phase I (day 1-60), selenite and tellurite were provided at 0.05 mM each, corresponding to 4 mg L^{-1} Se and 6 mg L^{-1} Te. In phase II (day 60-120), the selenite and tellurite concentrations were doubled to 0.1 mM each, i.e. 8 mg L^{-1} Se and 12 mg L^{-1} Te. The reactor was closed with a nylon rubber cork containing a vent which was connected to a suitable gas trap (0.5% HNO$_3$) for collection of the volatile fractions of Se and Te. Influent and effluent samples were collected every 48 h and stored at 4 °C until analysis.

8.2.5. Characterization of Se and Te nanostructures deposited in anaerobic granular sludge

At the end of the UASB reactor operation (day 120), selenite- and tellurite-reducing granular sludge was collected and air dried at 30 °C in order to analyze selenium and tellurium nanoparticles. The dried sludge was pulverized for analysis by Raman spectroscopy and P-XRD.

Micro-Raman spectroscopy (Renishaw inVia Raman microscope, United Kingdom) was used with a 2 μm diameter laser beam to focus onto the optical image of interest for recording Raman spectra. The excitation wavelength of 532 nm of the Nd-YAG laser was used for collecting Raman signals. P-XRD analysis was carried out with a Bruker D8 Advance using Cu Kα (1.54 Å) radiation in the range of 2θ = 20-90°. Since peak broadening was observed, the nanoparticles size (crystallite size) was determined using the Debye-Scherrer equation (Prasad et al. 2013):

$$\tau = \frac{K\lambda}{\beta \cos\theta} \tag{8.1}$$

where τ is the mean particle size (nm), K is a dimensionless shape factor (0.9), λ is the wavelength of the source used (nm), β is the full width at half maximum (FWHM) of the maximum intensity peak (radians) and θ is the corresponding Braggs angle (radians).

8.2.6. Extraction and characterisation of Se and Te nanostructures present in the anaerobic granular sludge

For the recovery of the Se and Te nanostructures from the granular sludge, a protocol used for extracting loosely bound extracellular polymeric substances (LB-EPS) was employed. At the end of the reactor run (day 120), the granular sludge biomass from the UASB reactor was washed with deionized water and the LB-EPS was extracted according to the procedure detailed by Zhao et al. (2015). The extract was characterized by TEM/STEM-EDS and ATR-IR spectroscopy. Electron microscopic imaging of the Se and Te nanostructures was done by

TEM/STEM-EDS on a JEOL JEM-2100F (HRP) using an acceleration voltage of 200.0 kV and a probe current of 1.0 nA. An ATR-IR (Thermo Scientific Nicolet iS 5 FT-IR spectrometer) with a reflection angle of 45° was used to determine the functional groups in the LB-EPS.

8.2.7. Analytical procedures

The influent and effluent samples from the UASB reactor were analyzed at 48 h intervals for COD, selenite, total selenium, Se(0), tellurite, total tellurium and Te(0). Effluent samples were processed by filtering through 0.45 μm cellulose acetate filters, followed by filtration through 0.1 μm cellulose acetate filters, to remove suspended microbial cells before measuring COD, selenite and tellurite. COD was measured using standard methods (APHA/AWWA/WEF 2012).

Selenite and tellurite concentrations in the influent and effluent were measured spectrophotometrically, respectively, using the ascorbic acid method (Mal et al. 2016a) and diethyldithiocarbamate (DDTC) method (Mal et al. 2017b). For selenite measurements, 2-mL filtered samples were added to 1 mL of 4 M hydrochloric acid and 2 mL of 1 M ascorbic acid (Mal et al. 2016a). The mixture was incubated at room temperature for 10 min prior to measuring absorbance at 500 nm on a UV-Vis spectrophotometer (Lambda-365, Perkin-Elmer). For measurements of tellurite, 0.5-mL filtered sample were added to 1.5 mL of 0.5 M Tris-HCl (pH 7.0) and 0.5 mL 10 mM DDTC reagent and the absorbance was measured at 340 nm (Mal et al. 2017b).

Total selenium and tellurium concentrations in the influent and effluent were analyzed using a graphite furnace atomic absorption spectrophotometer (Solaar AA Spectrometer GF95, Thermo Electrical) after digestion by 0.5% HNO_3 and suitable dilution. Samples were filtered with 0.1 μm cellulose acetate filters and the remaining concentration of selenium and tellurium was measured in the filtrates (Dessì et al. 2016; Mal et al. 2017b). The difference in the total selenium and tellurium content with remaining selenium and tellurium was considered to be Se(0) and Te(0), respectively.

8.2.8. Chemicals

Sodium selenite (99%) and potassium tellurite (90%) were obtained from Sigma-Aldrich (Steinheim, Germany). All other analytical-grade chemicals were purchased from Merck (Darmstadt, Germany). Stock solutions of 100 mM sodium selenite and 100 mM potassium tellurite were prepared in de-ionized water and stored at 4 °C.

8.3. Results

8.3.1. Selenite and tellurite reduction by anaerobic granular sludge

Figure 8.2a and 8.2b show the effect in batch assays of the sludge additions on the reduction of selenite and tellurite in wastewater medium. In the presence of 2 and 3 g wet weight of granular sludge, complete tellurite reduction was observed in 10 h. The tellurite reduction rate amounted to 6.7 and 4.0 µmoles of tellurite per g-VS per h respectively. Selenite was bio-reduced by 2 and 3 g wet weight of granules within 32 h and 24 h, which corresponds to the reduction rates of 2.0 and 1.7 µmoles selenite per g-VS per h, respectively.

In the presence of 1 mM tellurite, a complete selenite reduction by 2 and 3 g granules was, however, slower, requiring 50 and 35 h, respectively. The reduction rate with either biomass concentrations was 1.3 µmoles of selenite per g-VS per h (**Figure 8.2c**). In contrast, tellurite reduction was not affected by the presence of selenite (**Figure 8.2d**): reduction of 1 mM tellurite was completed within the first 10 h of the incubation at a reduction rate of 6.7 and 4.0 µmoles tellurite per g-VS per h for 2 and 3 g of granular sludge, respectively.

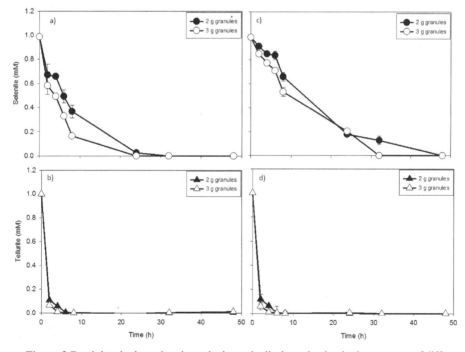

Figure 2 Batch incubations showing selenite and tellurite reduction in the presence of different initial biomass concentrations: (a, b) monocomponent incubations and (c, d) simultaneous reduction

of selenite and tellurite by anaerobic granular sludge showing the effect of tellurite on selenite bioreduction (c) and vice-versa (d). Note that the granule quantities are given in g wet weight.

8.3.2. UASB reactor performance

Tellurite removal

The influent tellurite and effluent Te (total Te and Te(0)) concentrations during UASB reactor operation are shown in **Figure 8.3a and 8.3b**. At the start-up of the UASB reactor, the tellurium removal efficiency was rather low, i.e. 57% (**Figure 8.3b**). It improved quickly to about 90% on day 3 of phase I of the UASB reactor operation and improved up to 96% by the end of phase I. Increasing the influent concentration of selenite and tellurite to 0.1 mM each in phase II affected the Te removal efficiency adversely, decreasing it from 97% to 81%. However, by the end of phase II and after 120 days of reactor operation, a Te removal efficiency of 98% was achieved. Tellurite removal coincided with the formation of elemental Te(0) in the UASB reactor (**Figure 8.3a**). In phase I, about 2% of the influent Te concentration (0.001 mmoles per liter) was released as Te(0) in the effluent. In phase II, the effluent Te(0) concentration increased up to 10% (0.01 mM) of the total influent concentration. By the end of the UASB reactor operation, the effluent Te(0) concentration was between 1-2% (0.001-0.002 mmoles per liter) of the influent Te concentration.

Selenite removal

Influent selenite and effluent Se (total Se and Se(0)) concentrations during UASB reactor operation are shown in **Figure 8.3c and 8.3d**. About 2% (0.001 mmoles per liter) of the total selenium added in the influent was present in the UASB reactor effluent during phase I mainly as Se(0) particles (**Figure 8.3c**). Doubling the selenite concentration in the UASB reactor influent decreased the selenium removal efficiency, the total selenium concentration in the effluent being 5% of the influent concentration at the start of phase II (**Figure 8.3d**). However, at the end of the reactor operation, improvement in the treatment efficiency was noticed, the Se(0) levels in the effluent being ≤ 2% of the influent selenite. **Figure 8.3d** shows the selenium removal efficiency of the UASB bioreactor as a function of time. A selenium removal efficiency of 95% was achieved in phase I of the UASB reactor operation. In phase II, the Se removal efficiency was initially only slightly affected and decreased to 87%, after which a stable selenium removal efficiency of 97% was achieved.

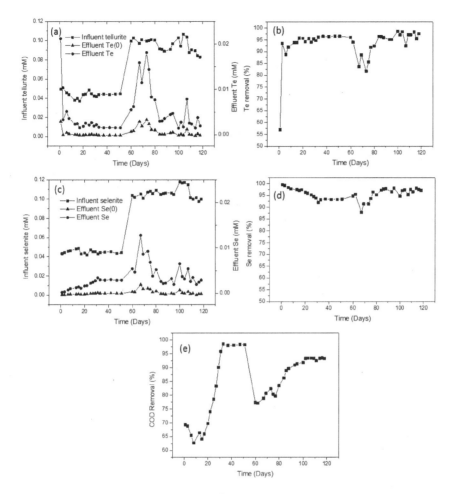

Figure 8.3 Removal of tellurite, selenite and COD from synthetic wastewater by the UASB reactor. (a) Tellurite concentration in influent and effluent, (b) Te removal efficiency, (c) selenite concentration in influent and effluent, (d) Se removal efficiency and (e) COD removal efficiency during 120 days of UASB reactor operation.

COD removal

In phase I, the COD removal efficiency was approximately 70% for the first 20 days of reactor operation, which gradually increased up to 98% by day 40 (**Figure 8.3e**). An immediate drop in the COD removal efficiency to 77% occurred at the start of phase II (day 60) upon doubling the influent selenite and tellurite concentrations to 0.1 mM each. However, a gradual improvement in COD removal efficiency up to 93% was noticed by the end of the reactor operation, possibly due to the acclimation of the granular sludge to the selenite and tellurite concentrations applied.

8.3.3. Characterization of immobilized Se, Te and Se-Te nanoparticles in the UASB granules

Characterization of the selenite and tellurite reducing granular sludge by Raman spectroscopy confirmed the formation of both Se(0) and Te(0) (**Figure 8.4a**). Two of the characteristic vibration peaks at 124 and 142 cm^{-1} were attributed to elemental Te(0) (Bonificio and Clarke, 2014; Mal et la. 2017b), while the peak at 235 cm^{-1} was assigned to trigonal-Se (t-Se) (Iovu et al. 2005). The broadening of the t-Se peak at 237 cm^{-1} depicts the A$_1$ mode of hexagonal selenium (Ren et al. 2004), while a split into several distinct peaks at ~208 and 171 cm^{-1} were also visible indicating presence of Se-Te mixed crystals (Geick et al. 1972).

The P-XRD spectra (**Figure 8.4b**) showed a good agreement with the Se and Te peaks when compared with the reference spectra of the standard database (Se-JCPDS No. 00-006-0362 and Te-JCPDS No. 36-1452) (Fu et al. 2015; Tao et al. 2009). The main diffraction peaks noticed in the spectra can be assigned to tellurium at (100), (101), (102) and (110) and to selenium at (100), (101), (110), (102) and (111). The peaks indexed in **Figure 8.4b** show that the selenium and tellurium nanoparticles are of a single phase and have a trigonal geometry. Since peak broadening was observed, the respective mean particle size of the Se and Te nanostructures for the Se (29.73°) and Te (27.7°) peak could be calculated using the Debye-Scherrer formula. It amounted to 52 and 37 nm, respectively.

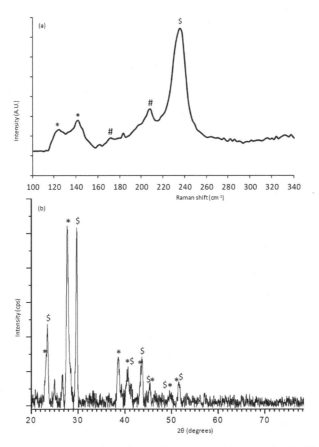

Figure 8.4 Spectroscopic characterisation of Se and Te nanoparticles associated with granular sludge performing simultaneous reduction of selenite and tellurite. (a) Raman spectra of anaerobic granular sludge treating synthetic wastewater containing selenite and tellurite. (*) show peaks assigned to pure elemental Te(0) at the expected positions of ~124 and 142 cm^{-1}. ($) indicates the peak for trigonal-Se at ~ 235 cm^{-1}. Peak broadening and splitting into several narrow distinct peaks are visible. (#) peaks at ~208 and ~ 171 cm^{-1} indicating the formation of Se/Te alloys, (b) X-ray diffraction pattern of the Se and Te nanoparticles found in the EPS extracted from selenite and tellurite reducing granular sludge. The peaks showing different intensities are identified as (*) and ($). (*) shows the peak position at 22.96°, 27.7°, 38.5° and 40.4° which indicates the presence of hexagonal Te, while ($) shows the peak position at 23.6°, 29.73°, 41.11°, 43.74° and 45.24° which indicates the presence of hexagonal Se. The maximum intensity for Te and Se was observed for peaks at 27.7° and 29.73°, respectively.

8.3.4. Recovery and characterization of Se and Te nanostructures from the UASB granules

Recovery of biogenic nanoparticles from UASB granules

Both Se and Te were found in the LB-EPS fraction, extracted from the granular sludge performing simultaneous reduction of selenite and tellurite. Total Se and Te concentrations amounted to 37.4 (± 1.3) and 40.2 (± 1.3) µg g^{-1} dry granular sludge, respectively.

Electron microscopic analysis of LB-EPS extracted from UASB granules

Clusters of irregular shaped nanoparticles embedded in a transparent layer of the LB-EPS were observed in the TEM images (**Figure 8.5a**). The size of these clusters ranged from 100 to 200 nm (**Figure 8.5a**). EDS analysis using the STEM mode confirmed the presence of peaks for both selenium and tellurium in the nanoparticle aggregates (**Figure 8.5b, c**).

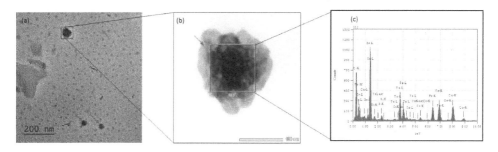

Figure 5 TEM and STEM images of the LB-EPS extracted from selenite and tellurite reducing granular sludge showing selenium and tellurium nanoparticles formed within the UASB granules. (a) Clusters of nanostructures, arrow points to a cluster of nanostructures, (b) arrow indicating EPS layer surrounding nanostructures, and STEM-EDS analysis of the selenium or tellurium nanoparticles show prominent peaks of the elements selenium and tellurium in EDS spectrum (c).

ATR-IR spectroscopy of LB-EPS extracted from UASB granules

The LB-EPS extracted from the granular sludge was analysed by ATR-IR spectroscopy (**Figure 8.6**). EPS consist of polysaccharides containing large numbers of hydroxyl groups, which exhibit a broad stretching peak around 3272 cm^{-1}. The presence of two peaks at 2922 cm^{-1} corresponds to methyl and methylene groups (Wang et al. 2010). The peak at 1629 cm^{-1} displays a strong absorption, corresponding to the amide stretch and C=N bonding of proteins and peptide amines, whereas the peak at 1530 cm^{-1} corresponds to N-H bending in amines. The peak at 1453 cm^{-1} corresponds to asymmetric deformation of CH$_3$ and CH$_2$ of proteins (Kumar et al. 2011; Wang et al. 2008). The peak at 1396 cm^{-1} can be assigned to the C=O stretch of the acid groups and C=O from COO$^-$ groups. In the fingerprint area (region below 1500 cm^{-1} where bands characterize the molecule as a whole), the peak at 1039 cm^{-1} was strong and can be attributed to the presence of polysaccharides (Wang et al. 2010).

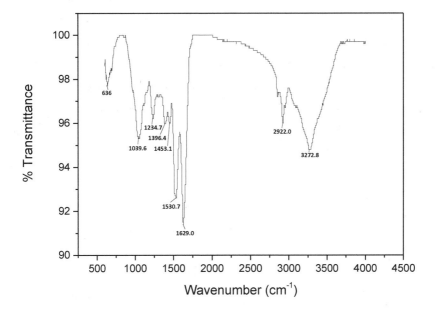

Figure 6 ATR-IR spectrum of the EPS extracted from selenite and tellurite reducing granular sludge. Peaks (cm^{-1}) in the spectrum represent the following groups: the C–H bend of the C-H bond adjacent to the carbon-carbon triple bond (636), polysaccharides (1039.6), protein amides (1234.7), acid groups (1396.4), CH$_3$- and CH$_2$- of proteins (1453.1), amines (1530.7), proteins and peptide amines (1629.0), methyl and methylene groups (2922.0) and hydroxyl groups (3272.8).

8.4. Discussion

8.4.1. Concomitant removal of selenite and tellurite by anaerobic granular sludge

This study demonstrated for the first time the simultaneous removal of selenite and tellurite by anaerobic granular sludge in a UASB reactor treating selenium and tellurium contaminated synthetic wastewater. Batch experiments showed that selenite reduction was faster with an increase in the amount of inoculum of anaerobic granular sludge (**Figure 8.2a**). They also revealed that tellurite reduction was faster than selenite reduction but was not affected by the inoculum concentration (**Figure 8.2b**). Interestingly, selenite bioreduction became slower in the presence of tellurite, however, tellurite bioreduction was not affected by the presence of selenite (**Figure 8.2c, d**). Slower reduction of selenite in the presence of tellurite could be attributed to the higher toxicity of tellurite to microorganisms (Turner et al. 2012) and interference in the reduction via detoxification. Tellurite reduction was unaffected in the presence of selenite, possibly due to its rapid removal and preference for reduction over selenite via detoxification mechanisms (e.g. glutathione dependent reduction). In contrast, Bajaj and Winter (2014) reported that selenite reduction by *Duganella violacienigra* was not affected by the presence of tellurite under aerobic conditions, while tellurite reduction increased 13-fold in the presence of selenite. Removal of both tellurite and selenite by the white-rot fungus *Phanerochaete chrysosporium* decreased by ~10 and ~15%, respectively, under aerobic conditions (Espinosa-Ortiz et al. 2017). Although, the exact interaction mechanisms of selenite and tellurite reduction are not clear, the results of the batch assays supported the development of a UASB bioreactor configuration to simultaneously remove selenium and tellurium oxyanions from wastewaters.

Speciation analysis during UASB reactor operation showed that >95% of selenite and tellurite were removed from the influent and immobilized as nanoparticles in the anaerobic granules in the UASB reactor, whereas around 2% of the Se and Te as Se(0) and Te(0) or Se-Te were present in the effluent (**Figure 8.3**). No significant change in the concentration of selenite or tellurite was observed in synthetic wastewater in the absence of granular sludge. Selenite and tellurite removal in the absence of an electron donor or with autoclaved anaerobic granular sludge was also very slow and not substantial. This confirms that anaerobic granular sludge is capable of reducing selenite or tellurite individually (Ramos-Ruiz et al. 2016). The batch experiments showed that the volatile Se and Te fractions were negligible compared to the total initial Se and Te concentration (data not shown). This was previously reported for the treatment of wastewater contaminated with Se (Dessì et al. 2016) or Te (Mal et al. 2017b) in the similar UASB reactor configuration as applied in this study. Most of the Se(0), Te(0) or Se-Te synthesized by the bioreduction of Se and Te oxyanions was immobilized in the granular sludge, resulting in the high Se and Te removal efficiencies. All previous work focused solely on either selenite or tellurite reduction in batch assays (Ramos-Ruiz et al. 2016) or continuous reactors (Mal et al. 2017b). The data (**Figure 8.2 and 8.3**) presented here show that microorganisms present in anaerobic granular sludge can perform simultaneous reduction and removal of both selenite and tellurite during batch and continuous UASB reactor operation at the applied loading rates and operational conditions. UASB reactors are thus a promising technology for simultaneous removal of both the metalloids from polluted effluents or groundwater and can be used for microbial synthesis of nanostructures of Se(0), Te(0) or Se-Te alloys.

The Te removal efficiency in the presence of selenite (**Figure 8.3b**) was comparable to that achieved in a recent study, where synthetic wastewater containing solely 10 mg L^{-1} tellurite was treated efficiently in a UASB reactor inoculated with the same inoculum as used in this study and operated under similar conditions (Mal et al. 2017b). A short acclimatisation period of only 2 days was required for the microorganisms to efficiently start tellurite and selenite reduction upon exposure to the oxyanions as well as upon doubling the oxyanion concentration in phase II (**Figure 8.3**). Acute toxicity associated with selenium and tellurium oxyanions may cause this delay in reduction (Mal et al. 2017c). Toxicity of selenite and tellurite on axenic bacterial cultures has been thoroughly investigated (Turner et al. 2012). The minimum inhibitory concentrations for selenite in bacterial cultures of *Escherichia coli*, *Staphylococcus aureus*, *Pseudomonas aeruginosa*, *P. flurescens*, *P. pseudoalcaligenes* and *Candida tropicalis* are 8.1, 16, 28, 8.7, 36.7 and 56 mM, respectively. In contrast, the minimum inhibitory concentrations for tellurite for these bacteria are much lower, i.e. 0.006, 0.18, 0.73, 0.03, 0.098 and >11 mM, respectively (Turner et al. 2012). The fungus *P. chrysosporium* showed the highest toxicity towards tellurite + selenite together, followed by selenite and the least toxicity towards tellurite (Espinosa-Ortiz et al. 2017). The mixed microbial community present in the granular sludge used in this study was susceptible to the high oxyanion concentrations (0.1 mM each) in the influent. The microbial community overcame the toxicity both in the batch incubations (**Figure 8.2**) and UASB reactors, when run with selenite (Dessì et al. 2016), tellurite (Mal et al. 2017b) or a mixture of both oxyanions (**Figure 8.3**).

Although the selenite and tellurite removal efficiency was ~97% for both oxyanions, the effluent still contained 0.002 mM of both elements as Se(0) and Te(0), which were represented by 0.16 mg L^{-1} and 0.28 mg L^{-1} of selenium and tellurium, respectively (**Figure 8.3**). Although a critical element in a wide-range of industrial applications, tellurium has not yet been considered a significant pollutant by the USEPA, which has not set its wastewater discharge limit. However, tellurium is toxic to living organisms even at low concentrations (Etezad et al. 2009). Reduction of this oxyanion to a non-toxic form, coupled with its recovery, is a promising approach to bioremediation of Te oxyanion-polluted wastewater.

The Te removal efficiency in the presence of selenite was better than in our previous study (Mal et al. 2017b). Mal et al. (2017b) found 0.2 and 1.6 mg total Te per liter of effluent when treating a synthetic wastewater containing 10 and 20 mg tellurite per liter, respectively, in a UASB reactor under similar conditions of operation, using the same anaerobic granular sludge as inoculum as used in the present study. Ramos-Ruiz et al. (2017) studied tellurium removal in a UASB reactor using ethanol as electron donor. Influent containing 10, 20 and 40 mg tellurite per liter was introduced into a UASB reactor in different phases at a HRT of 14.4 h. Tellurite was completely removed from the influent containing 10 and 20 mg L^{-1} Te (Ramos-Ruiz et al., 2017). The presence of Te(0) nanoparticles in the effluent was negligible. However, at tellurium concentrations of 40 mg L^{-1}, the Te removal efficiency decreased to 33%, which accounted for a Te effluent concentration of 26.8 mg L^{-1} by the end of the UASB reactor operation. This finding was attributed to the toxicity of Te to the anaerobic granular sludge (Ramoz-Ruiz et al. 2017).

Treatment of selenium containing wastewater has been widely investigated (Dessì et al. 2016; Lenz et al. 2008). The concentrations of the elemental Se in the treated water leaving the bioreactors were higher than those found in previous studies in which anaerobic granular sludge (Dessì et al. 2016; Gonzalez-Gil et al. 2016; Lenz et al. 2008) and the aerobic fungus *P. chrysosporium* (Espinosa-Ortiz et al. 2014) were used. However, it should be noted that the influent in the present study had a 10 and 2 times higher selenite concentration than that in the studies by Lenz et al. (2008) and Dessì et al. (2016), respectively. This could be one of the reasons for the higher Se(0) concentration in the effluent. The total Se concentration in the UASB reactor effluent was higher than that is permissible for environmental discharge. The USEPA has defined selenium as a "priority" pollutant, with a discharge limit in industrial wastewater of 5 µg L^{-1} (Nakamaru and Altansuvd 2014; Tan et al. 2016).

Selenium is often not the only pollutant present in the wastewater. It is often accompanied by oxyanions of its chalcogen counterpart, tellurium (Tan et al. 2016) as well as metals (Mal et al. 2016b). They can influence the selenite bioreduction mechanisms (intracellular versus extracellular reduction) or the presence of selenite may influence the removal of the other chalcogen oxyanions or metals, resulting in different Se(0)/Te(0) effluent concentrations. Incorporation of Te atoms into the Se-rings during biosynthesis of the Se(0) nanoparticles (and *vice versa*) could also influence the colloidal properties and settling characteristics of the nanoparticles. The release of Se-Te nanostructures from the granular sludge into the effluent could thus influence the Se/Te effluent concentration. Further studies are required to better understand the exact reduction mechanism (intracellular versus extracellular reduction) as well as removal mechanisms (e.g. biosorption, settling properties of colloidal nanoparticles) (Mal et al. 2016a) in multi-oxyanion contaminated wastewaters. Nevertheless, this study demonstrated simultaneous removal of selenite and tellurite with concomitant formation of Se(0), Te(0) and Se-Te nanostructures and that the removal of the oxyanions was consistent and stable during the 120 days of UASB reactor operation, even at high influent concentrations of both selenium and tellurium.

8.4.2. Characterization of biogenic Se(0), Te(0) and Se(0)-Te(0) nanostructures

Raman spectroscopy showed the presence of distinct peaks indicating the presence of pure metallic tellurium and trigonal selenium (**Figure 8.4a**) in the granular sludge of the UASB reactor in which the selenite and tellurite containing wastewater was treated. In nature, selenium occurs in different allotropic forms, including monoclinic, rhombohedral, trigonal and amorphous (Nagata et al. 1981), whereas the Te atoms occur in helices forming a hexagonal array (Brown and Forsyth, 1996). Chemically synthesized Se-Te alloys may form nanorods, nanowires or different anisotropic crystal structures (Fu et al. 2015). P-XRD analysis (**Figure 8.4b**) indicated that the Se(0) and Te(0) nanoparticles formed during reactor operation in the present study were of the trigonal-hexagonal crystal family. Similar characteristic peaks for Te(101) at 27.7° and for Se(101) at 29.73° were observed by Fu et al. (2015) during chemical synthesis of Se-Te alloy. Further evidence of the formation of both Se(0) and Te(0) nanocrystals and their retention in the granular sludge was observed in the characteristic diffraction peaks of crystalline Se(0)/Te(0) (**Figure 8.4b**). Complete reduction of selenite and tellurite to Se(0) and Te(0), respectively by the anaerobic granular sludge was further confirmed by the absence of characteristic peaks for Na_2SeO_3 or K_2TeO_3 by either Raman spectroscopy or P-XRD analysis (**Figure**).

In contrast, TEM images (**Figure 8.5a**) of the biogenic nanostructures recovered in the extracted LB-EPS show the presence of irregular structures in the form of clusters, suggesting the formation of anisotropic nanostructures8.4. STEM-EDS (**Figure 8.5b, c**) analysis of the biogenic nanostructures further confirmed the presence of Se(0) and Te(0) nanoparticles in clusters associated with LB-EPS extracted from UASB granules. They can be easily recovered by centrifugation (Mal et al. 2017a). Bajaj and Winter (2014) also reported the presence of extracellular, spherical nanoparticles of Se(0) and Te(0) after concurrent reduction of selenite and tellurite by *D. violacienigra*. Because of the isomorphic and anisotropic properties of Se and Te, they have many similar physical, chemical and crystalline properties. Although it is not clear from the EDS data whether it was Se(0), Te(0) or an Se-Te alloy, it is possible that the Se(0) and Te(0) atoms formed via microbial reduction interact and grew randomly to form mixed atom nanostructures (Gates et al. 2002; Mayers et al. 2001).

Raman spectroscopy clearly indicated the formation of new structural units (**Figure 8.4a**), possibly Se-Te alloy consisting of both Te and Se atoms (Geick et al. 1972; Tang et al. 2015). Notable broadening and splitting of the peaks suggests the inclusion of Te in the Se ring structure. The peak at ~ 208 cm^{-1} in the Raman spectra is assigned to vibration of neighbouring amorphous Se-Te bonds. The peak for the crystalline Se-Te appeared at ~ 171 cm^{-1} (Tang et al. 2015). Thus, formation of biogenic Se-Te structures was confirmed by the Raman spectroscopy. The crystal structure of the Se-Te nanoparticles, however, needs to be further characterized by X-ray absorption fine structure (XAFS) and X-ray absorption near edge structure (XANES) as performed for sulfur-selenium nanoparticles (Vogel et al., 2018), but this was beyond the scope of this study.

TEM images also revealed a distinct and translucent layer observed along with clusters of nanostructures (**Figure 8.5b**), which was most likely composed of the EPS produced by the anaerobic granules. The EPS matrix plays an important role in the capping and retention of either Se(0) (Jain et al. 2015) and Te(0) (Mal et al. 2017b). Biogenic selenium nanoparticles recovered from the wastewater were indeed coated with an EPS layer making them non-toxic and stable (Mal et al. 2017c). ATR-IR analysis (**Figure 6**) further confirmed that the functional groups in the LB-EPS were polysaccharides and proteins. The peaks representing the functional groups in the IR spectra were consistent with those shown by Jain et al. (2015) in an EPS study of an anaerobic selenite reducing granular sludge.

This study suggests that EPS plays a prominent role in 1) the retention of biogenic nanostructures of individual Se and Te nanoparticles, 2) the retention of biogenic Se-Te nanostructures and 3) capping of biogenic nanostructures to maintain their physiochemical stability. Based on the data, the proposed mechanism of formation of the Se, Te and Se-Te nanostructures by the anaerobic granular sludge (**Figure 8.7**) includes the following steps: 1) reduction of selenite to Se(0), 2) reduction of tellurite to Te(0), 3) formation of Se(0) and Te(0) nanostructures from Se(0) and Te(0) atoms, and 4) incorporation of Te atoms in Se rings and incorporation of Se atoms in Te rings that lead to the formation of Se-Te nanostructures.

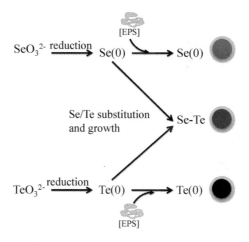

Figure 8.7 Proposed mechanism of Se(0), Te(0) and Se-Te nanostructures formed during the microbial reduction of selenite and tellurite.

8.4.3. Practical implications

This study presents, for the first time, possibilities for the biological synthesis of Se-Te nanostructures using anaerobic microorganisms. Superior properties such as electrical resistance and magnetoresistance makes Se-Te based alloys more applicable at an industrial scale rather than conventional pure Se and Te nanomaterials (Fu et al. 2015). Jian and Yadong (2003) reported that band parameters and transport properties of Se-Te crystalline systems can be easily modulated by changing the elemental composition during the chemical synthesis of Se-Te based nanostructures. This study showed these Se-Te nanostructures can also be synthesized using biosynthesis, as recently also reported for the biosynthesis of CdSe quantum dots (Mal et al. 2017a). The proof-of-concept demonstrated here needs to be further developed to fully characterize the biogenic Se-Te nanostructures, find technological applications, and up-scale the microbial manufacturing process to real waste streams.

Biological treatment of Se and Te containing wastewater is cost effective and environmentally friendly (Tan et al. 2016). UASB reactors allow entrapment of these critical elements in elemental form into the biomass which can be easily recovered from the bioreactor (**Figure** 3). The nanoparticles are mostly embedded in the LB-EPS of the anaerobic granules (**Figure** 4), making the separation of the nanoparticles from the granules easier: a simple procedure such as centrifugation, commonly used for LB-EPS extraction, can be used to recover the nanoparticles from the granular sludge of the UASB reactors. The same biomass can then be reused in the bioreactor, thus avoiding further acclimatization phases that are usually involved when using new granular sludge biomass in the UASB reactor operation.

References

APHA/AWWA/WEF (2012) Standard methods for the examination of water and wastewater, 22nd ed, Standard Methods. American Public Health Association, American Water Works Association, Water Environment Federation, Washington.

Baesman SM, Bullen TD, Dewald J, Zhang D, Curran S, Islam FS, Beveridge TJ, Oremland RS (2007) Formation of tellurium nanocrystals during anaerobic growth of bacteria that use Te oxyanions as respiratory electron acceptors. Appl Environ Microbiol 73:2135–2143.

Bajaj M, Winter J (2014) Se (IV) triggers faster Te (IV) reduction by soil isolates of heterotrophic aerobic bacteria: formation of extracellular SeTe nanospheres. Microb Cell Fact 13:1–10.

Bonificio WD, Clarke DR (2014) Bacterial recovery and recycling of tellurium from tellurium-containing compounds by *Pseudoalteromonas* sp. EPR3. J Appl Microbiol 117:1293–1304.

Brown PJ, Forsyth JB (1996) The crystal structure and optical activity of tellurium. Acta Crystallogr A52: 408–412.

Chasteen TG, Bentley R (2003) Biomethylation of selenium and tellurium. Chem Rev 103: 1–25.

Dessì P, Jain R, Singh S, Seder-Colomina M, van Hullebusch ED, Rene ER, Ahammad SZ, Carucci A, Lens PNL(2016) Effect of temperature on selenium removal from wastewater by UASB reactors. Water Res 94:146–154.

Dhillon KS, Dhillon SK (2003) Distribution and management of seleniferous soils.In: Sparks D. (ed) Advances in Agronomy, Academic Press. pp 119–184.

Espinosa-Ortiz EJ, Gonzalez-Gil G, Saikaly PE, van Hullebusch ED, Lens PNL (2014) Effects of selenium oxyanions on the white-rot fungus *Phanerochaete chrysosporium*. Appl Microbiol Biotechnol 99:2405–2418.

Espinosa-Ortiz EJ, Rene ER, Guyot F, van Hullebusch ED, Lens PNL (2017) Biomineralization of tellurium and selenium-tellurium nanoparticles by the white-rot fungus *Phanerochaete chrysosporium*. Int Biodeterior Biodegradation 124:258-266.

Etezad SM, Khajeh K, Soudi M, Ghazvini PTM, Dabirmanesh B (2009) Evidence on the presence of two distinct enzymes responsible for the reduction of selenate and tellurite in *Bacillus* sp. STG-83. Enzyme Microb Technol 45:1–6.

Fu S, Cai K, Wu L, Han H (2015) One-step synthesis of high-quality homogenous Te/Se alloy nanorods with various morphologies. CrystEngComm 17:3243–3250.

Gates B, Mayers B, Cattle B, Xia Y (2002) Synthesis and characterization of uniform nanowires of trigonal selenium. Adv Funct Mater 12:219–227.

Geick R, Steigmeter EF, Auderset H (1972) Raman effect in selenium-tellurium mixed crystasls. Phys Stat Sol 54:623-630.

George MW (2003) Selenium and tellurium. In: U.S. Geological Suvery Minerals Yearbook. pp 1–8.

Gonzalez-Gil G, Lens PNL, Saikaly PE (2016) Selenite reduction by anaerobic microbial aggregates: Microbial community structure, and proteins associated to the produced selenium spheres. Front Microbiol 7:1–14.

Gonzalez-Gil G, Seghezzo L, Lettinga G, Kleerebezem R (2001) Kinetics and mass-transfer phenomena in anaerobic granular sludge. Biotechnol Bioeng 73:125–134.

Hunter WJ (2014) A *Rhizobium selenitireducens* protein showing selenite reductase activity. Curr Microbiol 68:311–316.

Iovu MS, Kamitsos EI, Varsamis CPE, Boolchand P, Popescu M (2005) Raman spectra of As_xSe_{100-x} and $As_{40}Se_{60}$ glasses doped with metals. Chalcogenide Lett 2:21–25.

Jain R, Jordan N, Weiss S, Foerstendorf H, Heim K, Kacker R, Hubner R, Kramer H, Van Hullebusch ED, Farges F, Lens PNL (2015) Extracellular polymeric substances govern the surface charge of biogenic elemental selenium nanoparticles. Environ Sci Technol 49:1713–1720.

Jian X, Yadong L (2003) Solution route to Se_xTe_{1-x}/Te/Se_xTe_{1-x} heterojunction nanorods. Mater Chem Phys 82:515–519.

Jorgenson JD (2002) Selenium and tellurium, In: U.S. Geological Suvery Minerals Yearbook. pp 1–7.

Kagami T, Fudemoto A, Fujimoto N, Notaguchi E, Kanzaki M, Kuroda M, Soda S, Yamashita M, Ike M (2012) Isolation and characterization of bacteria capable of reducing tellurium oxyanions to insoluble elemental tellurium for tellurium recovery from wastewater. Waste Biomass Valorization 3:409–418.

Kumar MA, Anandapandian KTK, Parthiban K (2011) Production and characterization of exopolysaccharides (EPS) from biofilm forming marine bacterium. Brazilian Arch Biol Technol 54:259–265.

Lenz M, van Hullebusch ED, Hommes G, Corvini PFX, Lens PNL (2008) Selenate removal in methanogenic and sulfate-reducing upflow anaerobic sludge bed reactors. Water Res 42:2184–2194.

Li D-B, Cheng Y-Y, Wu C, Li W-W, Li N, Yang Z-C, Tong Z-H, Yu H-Q (2014) Selenite reduction by *Shewanella oneidensis* MR-1 is mediated by fumarate reductase in periplasm. Sci Rep 4:3735.

Mal J, Nancharaiah YV, Bera S, Maheshwari N, van Hullebusch ED, Lens PNL (2017a) Biosynthesis of CdSe nanoparticles by anaerobic granular sludge. Environ Sci Nano 4:824–833.

Mal J, Nancharaiah YV, van Hullebusch ED, Lens PNL (2016a) Effect of heavy metal co-contaminants on selenite bioreduction by anaerobic granular sludge. Bioresour Technol 206:1–8.

Mal J, Nancharaiah YV, Maheshwari N, van Hullebusch ED, Lens PNL (2017b) Continuous removal and recovery of tellurium in an upflow anaerobic granular sludge bed reactor. J Hazard Mater 327:79–88.

Mal J, Nancharaiah YV, van Hullebusch ED, Lens PNL (2016b) Metal chalcogenide quantum dots: biotechnological synthesis and applications. RSC Adv 6:41477–41495.

Mal J, Veneman WJ, Nancharaiah YV, van Hullebusch ED, Peijnenburg WJGM, Vijver MG, Lens PNL (2017c) A comparison of fate and toxicity of selenite, biogenically and chemically synthesized selenium nanoparticles to to zebrafish (*Danio rerio*) embryogenesis. Nanotoxicology 1–11.

Mayers B, Gates B, Yin Y, Xia Y (2001) Large-scale synthesis of monodisperse nanorods of Se/Te alloys through a homogeneous nucleation and solution growth process. Adv Mater 13:1380–1384.

Nagata K, Ishibashi K, Miyamoto Y (1981) Raman and infrared spectra of rhombohedral selenium. Jpn J Appl Phys 20:463–469.

Nakamaru YM, Altansuvd J (2014) Speciation and bioavailability of selenium and antimony in non-flooded and wetland soils: A review. Chemosphere 111:366–371.

Nancharaiah YV, Mohan SV, Lens PNL (2016) Biological and bioelectrochemical recovery of critical and scarce metals. Trends Biotechnol 34(2):137-155.

Pat-Espadas AM, Field JA, Razo-Flores E, Cervantes FJ, Sierra-Alvarez R (2016) Continuous removal and recovery of palladium in an upflow anaerobic granular sludge bed (UASB) reactor. J Chem Technol Biotechnol 91:1183–1189.

Perkins WT (2011) Extreme selenium and tellurium contamination in soils - An eighty year-old industrial legacy surrounding a Ni refinery in the Swansea Valley. Sci. Total Environ. 412–413:162–169.

Prasad KS, Patel H, Patel T, Patel K, Selvaraj K (2013) Biosynthesis of Se nanoparticles and its effect on UV-induced DNA damage. Colloids Surf B Biointerfaces 103:261–266.

Rajwade JM, Paknikar KM (2003) Bioreduction of tellurite to elemental tellurium by *Pseudomonas mendocina* MCM B-180 and its practical application. Hydrometallurgy 71:243–248.

Ramos-Ruiz A, Field JA, Wilkening JV, Sierra-Alvarez R (2016) Recovery of elemental tellurium nanoparticles by the reduction of tellurium oxyanions in a methanogenic microbial consortium. Environ Sci Technol 50:1492–1500.

Ramos-Ruiz A, Sesma-Martin J, Sierra-Alvarez R, Field JA (2017) Continuous reduction of tellurite to recoverable tellurium nanoparticles using an upflow anaerobic sludge bed (UASB) reactor. Water Res 108:189–196.

Ren L, Zhang H, Tan P, Chen Y, Zhang Z, Chang Y, Xu F, Yang F, Yu D (2004) Hexagonal selenium nanowires synthesized via vapor-phase growth. J Phys Chem B 108:4627-4630.

Spinks SC, Parnell J, Bellis D, Still J (2016) Remobilization and mineralization of selenium-tellurium in metamorphosed red beds: Evidence from the Munster Basin, Ireland. Ore Geol Rev 72:114–127.

Staicu LC, Ackerson CJ, Cornelis P, Ye L, Berendsen RL, Hunter WJ, Noblitt SD, Henry CS, Cappa JJ, Montenieri RL, Wong AO, Musilova L, Sura-de Jong M, van Hullebusch ED, Lens PNL, Reynolds RJB, Pilon-Smits EAH (2015) *Pseudomonas moraviensis* subsp. stanleyae, a bacterial endophyte of hyperaccumulator *Stanleya pinnata*, is capable of efficient selenite reduction to elemental selenium under aerobic conditions. J Appl Microbiol 119:400–410.

Stams AJM, Grolle KCF, Frijters, CTMJ, Van Lier JB (1992) Enrichment of thermophilic propionate-oxidizing bacteria in syntrophy with *Methanobacterium thermoautotrophicum* or *Methanobacterium thermoformicicum*. Appl Environ Microbiol 58:346–352.

Tan LC, Nancharaiah YV, van Hullebusch ED, Lens PNL (2016) Selenium: Environmental significance, pollution, and biological treatment technologies. Biotechnol Adv 34:886–907.

Tang G, Qian Q, Wen X, Chen X, Liu W, Sun M, Yang Z (2015) Reactive molten core fabrication of glass-clad $Se_{0.8}Te_{0.2}$ semiconductor core optical fibers. Opt Express 23:23624–23633.

Tao H, Shan X, Yu D, Liu H, Qin D, Cao Y (2009) Solution grown Se/Te nanowires:

Nucleation, evolution, and the role of triganol Te seeds. Nanoscale Res Lett 4:963–970.

Turner RJ, Borghese R, Zannoni D (2012) Microbial processing of tellurium as a tool in biotechnology. Biotechnol Adv 30:954–963.

Vogel M, Fischer S, Maffert A, Hübner R, Scheinost AC, Franzen C, Steudtner R (2018) Biotransformation and detoxification of selenite by microbial biogenesis of selenium-sulfur nanoparticles. J. Haz. Mat. 344:749–757.

Wang Y, Ahmed Z, Feng W, Li C, Song S (2008) Physicochemical properties of exopolysaccharide produced by *Lactobacillus kefiranofaciens* ZW3 isolated from Tibet kefir. Int J Biol Macromol 43:283–288.

Wang Y, Li C, Liu P, Ahmed Z, Xiao P, Bai X (2010) Physical characterization of exopolysaccharide produced by *Lactobacillus plantarum* KF5 isolated from Tibet Kefir. Carbohydr Polym 82:895–903.

Yoon BM, Shim SC, Pyun HC, Lee DS (1990) Hydride generation atomic absorption determination of tellurium species in environmental samples with *in situ* concentration in a graphite furnace. Anal. Sci. 6:561–566.

Zhao W, Yang S, Huang Q, Cai P (2015) Bacterial cell surface properties: Role of loosely bound extracellular polymeric substances (LB-EPS). Colloids Surfaces B Biointerfaces 128:600–607.

CHAPTER 9

Discussion, Conclusion and Perspectives

9.1. General discussion

The importance of environmental selenium research is increasingly recognized during the last decade (Nancharaiah and Lens, 2015b; Tan et al., 2016). However, the concerns about Se toxicity began in 1930, when symptoms for alkali disease and blind staggers were observed in livestock grazing on grass grown on Se-enriched soil in South Dakota (Tinggi, 2003). While Se deficiency was brought to the forefront in 1960's with identification of a peculiar heart muscle disease symptom, called Keshan's disease in China (Chen, 2012). Both these Se deficiency and toxicity disorders are the ramifications of the Se bioavailability in the respective soils of the regions (China and South Dakota respectively). Several studies are being carried out to resolve the issue of Se imbalance across the globe, where efforts are being made to remove Se from the seleniferous regions (Bañuelos et al., 2005; Lindblom et al., 2014; Sura-de Jong et al., 2015) and fortify Se-deficient soils with organic and inorganic Se compounds (Bañuelos et al., 2015; Lyons, 2010). The major aim of this thesis was to develop a technology for the removal as well as recovery of Se from seleniferous soil. Se is a scarce and critical element (Nancharaiah et al., 2016) and its recovery from soil may assist not only in the clean-up of seleniferous soil, but also recovered Se may complement as a raw material for its wide range of applications from healthcare to electronic industries.

For this research, the seleniferous soil characteristics, Se migration pattern in soil and soil washing to achieve maximum removal of Se were determined (**Chapter 3**). Biological treatment of Se-containing soil leachate was optimized using a continuous bioreactor and phytoremediation by aquatic plants (**Chapter 4 and 5**). Biological reduction of selenium oxyanions under different microbial respiration conditions and bioreactor configurations was explored for removing Se from simulated artificial soil leachate (**Chapter 6, 7 and 8**). Bioreduction of selenium oxyanions under aerobic and anaerobic conditions was investigated using different inocula (pure culture of *D. lacustris*, marine lake sediment and anaerobic granular sludge). Under anaerobic conditions, the effect of different electron donors (methane and lactate) and electron acceptors as co-contaminants (tellurite) on selenium oxyanion reduction was studied. This **chapter 9** provides a thorough insight into the current status of *ex situ* bioremediation of seleniferous soils along with the environmental and economic feasibility of the process. The overview of the results for each research objective and future perspectives of these technologies based on the findings of the present thesis are described in **Figure 9.1**.

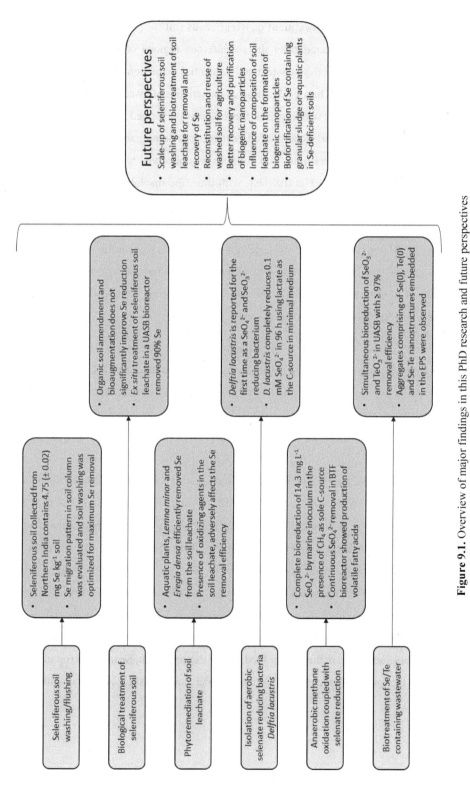

Figure 9.1. Overview of major findings in this PhD research and future perspectives

Future perspectives

- Scale-up of seleniferous soil washing and biotreatment of soil leachate for removal and recovery of Se
- Reconstitution and reuse of washed soil for agriculture
- Better recovery and purification of biogenic nanoparticles
- Influence of composition of soil leachate on the formation of biogenic nanoparticles
- Biofortification of Se containing granular sludge or aquatic plants in Se-deficient soils

---

- Organic soil amendment and bioaugmentation does not significantly improve Se reduction
- *Ex situ* treatment of seleniferous soil leachate in a UASB bioreactor removed 90% Se

- *Delftia lacustris* is reported for the first time as a SeO_4^{2-} and SeO_3^{2-} reducing bacterium
- *D. lacustris* completely reduces 0.1 mM SeO_4^{2-} in 96 h using lactate as the C-source in minimal medium

- Simultaneous bioreduction of SeO_3^{2-} and TeO_3^{2-} in UASB with ≥ 97% removal efficiency
- Aggregates comprising of Se(0), Te(0) and Se-Te nanostructures embedded in the EPS were observed

---

- Seleniferous soil collected from Northern India contains 4.75 (± 0.02) mg Se kg^{-1} soil
- Se migration pattern in soil column was evaluated and soil washing was optimized for maximum Se removal

- Aquatic plants, *Lemna minor* and *Eregia densa* efficiently removed Se from the soil leachate
- Presence of oxidizing agents in the soil leachate, adversely affects the Se removal efficiency

- Complete bioreduction of 14.3 mg L^{-1} SeO_4^{2-} by marine inoculum in the presence of CH_4 as sole C-source
- Continuous SeO_4^{2-} removal in BTF bioreactor showed production of volatile fatty acids

---

- Seleniferous soil washing/flushing
- Biological treatment of seleniferous soil
- Phytoremediation of soil leachate
- Isolation of aerobic selenate reducing bacteria *Delftia lacustris*
- Anaerobic methane oxidation coupled with selenate reduction
- Biotreatment of Se/Te containing wastewater

9.2. Selenium removal and biotreatment of seleniferous soil

In this research, soil was collected from the seleniferous regions of Punjab (India). The physico-chemical parameters, total Se content and sequential extraction of Se in the soil were characterized and presented in **Chapter 3**. These characteristics allowed to further design the experiments for soil flushing and washing. Soil flushing was performed to assess Se migration in a soil column by simulating artificial rainfall or irrigation. Se migration and accumulation from the upper to the lower layers in the soil columns was observed suggesting reduction of soluble Se to insoluble Se forms. Nevertheless, with time, the insoluble Se fraction may be slowly re-oxidised to soluble Se forms and contaminate groundwater. Soil washing was optimized in order to achieve a maximum removal of bioavailable forms of Se from the seleniferous soil. Biological treatment of the Se-rich soil leachate was further evaluated using microbial reduction in a continuous bioreactor (**Chapter 4**) and by phytoremediation using aquatic plants (**Chapter 5**). A schematic approach of the washing of seleniferous soil and biotreatment of the Se-rich soil leachate is shown in **Figure 9.2**.

Experiments were designed to optimize the biotreatment of seleniferous soil *in situ* and *ex situ* (**Chapter 4**). *In situ* microcosms were set up in order to evaluate the effect of the organic amendment and bioaugmentation on reduction of Se in the soil. However, organic amendment and bioaugmentation exhibited similar Se removal performance as those of the control setup without any amendment or bioaugmentation. This suggested that under ideal environmental conditions, the soil indigenous microbial population and organic content available in the soil were sufficient to achieve Se reduction *in situ*. Flury et al. (1997) also did not find major differences between the control and tests in a field study that attempted volatilization of Se from Kesterson reservoir (California, USA). But, the drawback of *in situ* method is that it only converts the bioavailable Se into insoluble Se forms, where the risk of its re-oxidation to soluble Se forms cannot be ruled out. Recovery of Se from the soil was achieved by implementing a novel *ex situ* approach for seleniferous soil bioremediation. The soil leachate from the seleniferous soil (leachate preparation presented in **Chapter 3**) was treated in an upflow anaerobic sludge bed (UASB) reactor, while gradually decreasing and finally excluding the C-source supplementation in the reactor feed. Using this approach, Se removal (up to 90%) and retention in the form of biogenic Se(0) in the granular sludge of the UASB was achieved.

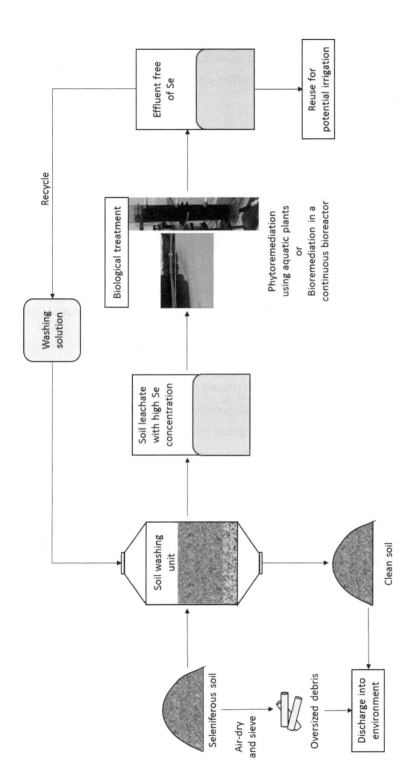

Figure 9.2. Schematic of the *ex situ* remediation approach showing washing of seleniferous soils and biotreatment of the Se-rich soil leachate adopted in this study

Along with the granular sludge, the indigenous soil microorganisms and the organic matter, extracted into the leachate during soil washing, play an important role to achieve efficient Se removal in the UASB reactor. This technology offers a low-cost and effective solution for treatment of seleniferous soils.

Oxidising agents were found to be most efficient in extracting Se from the studied seleniferous soil (**Chapter 3**). Phytoremediation of Se-rich leachate containing residual oxidising agents ($K_2S_2O_8$ and $KMnO_4$) was evaluated using the aquatic plants such as *Lemna minor* and *Eregia densa* and presented in **Chapter 5**. *L. minor* was found to be more efficient in removing Se from soil leachate and Se accumulation than *E. densa*. However, the presence of residues of the oxidising agents in the Se-rich effluent adversely affected not only the Se removal efficiency but also plant growth in both the plants. Although application of oxidising agents is promising for effective removal of Se from the seleniferous soils (**Chapter 3**), the use of chemical agents must be discouraged, as it might aggravate the post-treatment operations for both treated soil and the leachate.

9.3. Biological treatment by microbial reduction

Microbial reduction of soluble selenium oxyanions (selenate and selenite) to insoluble and non-reactive elemental selenium is the best available cost effective and eco-friendly option (Nancharaiah and Lens, 2015a). A part of this research, *Delftia lacustris* was reported for the first time as a selenate and selenite reducing bacterium (**Chapter 6**). Interestingly, this bacterium is capable of tolerating and reducing high concentrations of selenium oxyanions under aerobic conditions in minimal salt medium. Optimum conditions for reducing selenate and selenite by *D. lacustris* were determined. This bacterium not only reduces Se oxyanions to elemental selenium, but also produces seleno-ester compounds (organo-Se compound) depending on the initial concentrations of Se oxyanion in the medium. The results show that this organism has a unique capability to reduce selenate under aerobic conditions through a novel Se metabolism which is yet to be deciphered.

The effluent discharge from mining industries and coal fired power plants have resulted in large-scale Se deposition in marine sediments (Ellwood et al., 2016). In this context, anaerobic bioreduction of selenate to elemental Se by marine lake sediment in the presence of methane as a sole electron donor was investigated (**Chapter 7**). Most of the research on bioreduction of Se oxyanions utilize expensive C-sources such as lactate (Dessì et al., 2016; Mal et al., 2016) or glucose (Espinosa-Ortiz et al., 2015). While few studies explore the possibility of using inexpensive electron donors like methane (Lai et al., 2016), as sole electron donor for bioreduction of Se oxyanions. In this study, complete bioreduction of selenate was observed in serum bottles under high pressure conditions and in a biotrickling filter (BTF) reactor. However, due to the slow growing nature of the inoculum (Bhattarai et al., 2017), the operation time to achieve complete selenate removal from the medium was significantly higher than that of other studies that provide lactate as electron donor (Dessì et al., 2016; Mal et al., 2016). During the BTF reactor operation, volatile fatty acid (acetate and propionate) production, followed by their consumption was observed, possibly due to anaerobic oxidation of methane to form acetate and propionate (Lai et al., 2016). Further analysis of the microbial community from the sediment enrichments would further improve the understanding of the biochemical pathways on volatile fatty acid (acetate and propionate) production, utilization and selenate reduction coupled with methane oxidation.

Anthropogenic activities such as mining and refinery industries can lead to such contaminated soil-water environments containing both Se and Te (Perkins, 2011). Recovery of Se and Te as by-products from copper mining industries focusses on anode slime which contains high concentrations of Se and Te (Jorgenson, 2002). Both Se and Te are considered as critical elements because their wide range of applications and increasing demand in the electronic industries (Nancharaiah et al., 2016; Ramos-Ruiz et al., 2016). In this context, effect of metalloid co-contaminants (Te) on Se removal and simultaneous removal of both Se and Te was investigated (**Chapter 8**). This work demonstrated for the first time that simultaneous removal of both selenite and tellurite from synthetic wastewater by anaerobic granular sludge was sustainable in a lab-scale UASB reactor. Microbial transformations converted both toxic oxyanions (selenite and tellurite) to their respective elemental forms (Se(0) and Te(0)), which were entrapped in the EPS matrix of the granular sludge, easing the recovery of these critical elements. Characterization revealed that formation of biogenic Se(0)-Te(0) nanostructures during simultaneous reduction of Se and Te oxyanions. Formation, entrapment and characterization of these biogenic Se(0), Te(0) and Se(0)-Te(0) nanostructures were characterized using high-end microscopic and spectroscopic techniques. The overall results presented in this PhD thesis are novel and will advance the field of bioremediation and coupled resource recovery.

9.4. Future perspectives

In this research, a proof-of context was developed at lab scale for *ex situ* treatment of seleniferous soil. However, large scale operation of *ex situ* soil washing of thousands of hectares of agricultural land (seleniferous soil) (Dhillon and Dhillon, 2003) may pose the problem of implementation of this techniques in terms of cost. On the other hand, large-scale recovery of biogenic Se nanoparticles may help sustain economically the bioremediation process. However, other components in the soil leachate may influence the production and composition of the biogenic Se nanostructures. Along with biogenic Se(0), possibility of formation of chalcogenide alloys (Se-Te, Se-S) or metal chalcogenide nanoparticles (CdSe, CuSe) in the continuous bioreactor systems cannot be ruled out. Detailed analysis of the EPS of the granular sludge using high-end techniques such as Transmission Electron Microscopy coupled with Energy-dispersive X-ray spectroscopy (TEM-EDS), X-ray photoelectron spectroscopy (XPS) and Raman spectroscopy may help determine the structure and composition of the biogenic selenium nanoparticles. The treated effluent free of organic matter and Se may be discharged or recycled for soil washing, as proposed in **Figure 9.2**.

Prior to reconstitution of washed soil, suitability of the soil to be discharged in the environment should be determined. Unlike soil flushing, continuous agitation during soil washing disturbs the layers present in the soil column naturally, which might affect the soil characteristics. In addition, after soil washing, the soil characteristics and the bioavailability of Se might change in the course of time depending on the soil environmental conditions. Long term experiments in order to determine the physico-chemical characteristics of the washed soil, as well as alterations in the Se fractions during sequential extraction must be performed. Fertility of the soil may be checked using germination tests. Pot experiments to compare plant growth and Se uptake in seleniferous soil and washed seleniferous soil may be performed. On the other hand, heap flushing for Se extraction from soil may be practiced using a shallow layer of soil and slow percolation of the washing solution through the soil column for maximum Se extraction.

Extraction of selenium from seleniferous soil by soil washing and its accumulation in anaerobic granular sludge and aquatic plants helps easier recovery of Se from soil. These granular sludge or aquatic plants may be used for biofortification of selenium in crops. In addition, Se-containing aquatic plants can be considered for developing a dietary selenium supplement. Similar application of Se hyperaccumulator plants has been proposed to compensate the effects of selenium deficiency in selenium deficient regions (Moreno et al., 2013; Yasin et al., 2014). Several studies (Curtin et al., 2006; Lyons, 2010) have investigated the effect of addition of inorganic-Se (selenate and selenite) containing fertilizers for selenium biofortification in food crops in selenium deficient countries such as the UK and New Zealand. In Finland, selenium deficient soils have led to a low selenium status in humans and animals. A nation-wide strategy was adopted on application of multimineral fertilizers on agricultural soils to improve selenium dietary intake (Parkman et al., 2002).

Maintaining accurate and low selenium dosage in agricultural land can be difficult owing to diverse cropping systems and selenium losses due to leaching and volatilization. Also, selenium being a scarce resource with very low recovery potential when applying foliar fertiliser, amendment with organic-Se such as Se-enriched hyperaccumulator plant material to Se-deficient soils may prove as a better fertilizer alternative. Bañuelos et al. (2015) concluded that amending soils with organic selenium sources such as biomass of the hyperaccumulator *S. pinnata* is useful for enriching food crops, such as broccoli and carrots, with organic-Se in selenium deficient regions of the world. Bañuelos et al. (2000) suggested that Canola (*Brassica napus*) grown as a selected plant species for field phytoremediation of Se-contaminated soils may be harvested and utilized as Se-enriched forage for marginally Se-deficient lambs and cows to help meet their normal selenium intake requirements.

Se in the seleniferous soil collected for this research is of lithogenic origin and is transported via Se-rich sediments by seasonal rivulets from sub-Himalayan ranges called the Shiwalik hills in the north of Punjab (India) (Dhillon and Dhillon, 2009). Intensive irrigation on the local agricultural lands during the last few decades has increased the Se deposition in the soil and has led to accumulation of up to 1.4 kg Se per hectare every year (Dhillon and Dhillon, 2003). In order to avoid further contamination of these agricultural soils via rivulets containing Se-rich sediments, storage and sedimentation of the irrigation water may be practiced prior to irrigation on the agricultural fields. The Se-rich sediments may then be extracted and treated using a suitable procedure.

References

Bañuelos, G.S., Arroyo, I., Pickering, I.J., Yang, S.I., Freeman, J.L., 2015. Selenium biofortification of broccoli and carrots grown in soil amended with Se-enriched hyperaccumulator *Stanleya pinnata*. Food Chem. 166, 603–8.

Bañuelos, G.S., Lin, Z.Q., Arroyo, I., Terry, N., 2005. Selenium volatilization in vegetated agricultural drainage sediment from the San Luis Drain, Central California. Chemosphere 60, 1203–1213.

Bañuelos, G.S., Mayland, H.F., 2000. Absorption and distribution of selenium in animals consuming canola grown for selenium phytoremediation. Ecotoxicol. Environ. Saf. 46, 322–8.

Bhattarai, S., Cassarini, C., Naangmenyele, Z., Rene, E.R., 2017. Microbial sulfate-reducing activities in anoxic sediment from Marine Lake Grevelingen : screening of electron donors and acceptors. Limnology. doi:10.1007/s10201-017-0516-0

Chen, J., 2012. An original discovery: Selenium deficiency and Keshan disease (an endemic heart disease). Asia Pac. J. Clin. Nutr. 21, 320–326.

Curtin, D., Hanson, R., Lindley, T.N., Butler, R.C., 2006. Selenium concentration in wheat (*Triticum aestivum*) grain as influenced by method, rate, and timing of sodium selenate application. New Zeal. J. Crop Hortic. Sci. 34, 329–339.

Dessì, P., Jain, R., Singh, S., Seder-Colomina, M., van Hullebusch, E.D., Rene, E.R., Ahammad, S.Z., Carucci, A., Lens, P.N.L., 2016. Effect of temperature on selenium removal from wastewater by UASB reactors. Water Res. 94, 146–154.

Dhillon, K.S., Dhillon, S.K., 2003. Distribution and management of seleniferous soils, in: Advances in Agronomy. pp. 119–184.

Dhillon, K.S., Dhillon, S.K., 2003. Quality of underground water and its contribution towards selenium enrichment of the soil-plant system for a seleniferous region of northwest India, in: Journal of Hydrology. pp. 120–130.

Dhillon, S.K., Dhillon, K.S., 2009. Phytoremediation of selenium-contaminated soils: The efficiency of different cropping systems. Soil Use Manag. 25, 441–453.

Ellwood, M.J., Schneider, L., Potts, J., Batley, G.E., Floyd, J., Maher, W.A., 2016. Volatile selenium fluxes from selenium-contaminated sediments in an Australian coastal lake. Environ. Chem. 13, 68–75.

Espinosa-Ortiz, E., Rene, E.R., van Hullebusch, E.D., Lens, P.N.L., 2015. Removal of selenite From wastewater in a *Phanerochaete chrysosporium* pellet based fungal bioreactor. Int. Biodeterior. Biodegradation 102, 361–369.

Flury, M., Frankenberger Jr., W.T., Jury, W.A., 1997. Long-term depletion of selenium from Kesterson dewatered sediments 198, 259–270.

Jorgenson, J.D., 2002. Selenium and tellurium, in: U.S. Geological Suvery Minerals Yearbook. pp. 1–7.

Lai, C.-Y., Wen, L.-L., Shi, L.-D., Zhao, K.-K., Wang, Y.-Q., Yang, X., Rittmann, B.E., Zhou, C., Tang, Y., Zheng, P., Zhao, H.-P., 2016. Selenate and nitrate bioreductions using methane as the electron donor in a membrane biofilm reactor. Environ. Sci. Technol. 50, 10179–86.

Lindblom, S.D., Fakra, S.C., Landon, J., Schulz, P., Tracy, B., Pilon-Smits, E. A. H., 2014. Inoculation of selenium hyperaccumulator *Stanleya pinnata* and related non-accumulator

Stanleya elata with hyperaccumulator rhizosphere fungi - investigation of effects on Se accumulation and speciation. Physiol. Plant. 150, 107–18.

Lyons, G., 2010. Selenium in cereals: Improving the efficiency of agronomic biofortification in the UK. Plant Soil 332, 1–4.

Mal, J., Nancharaiah, Y. V., van Hullebusch, E.D., Lens, P.N.L., 2016. Effect of heavy metal co-contaminants on selenite bioreduction by anaerobic granular sludge. Bioresour. Technol. 206, 1–8.

Moreno, R.G., Burdock, R., Cruz, M., Álvarez, D., Crawford, J.W., 2013. Managing the selenium content in soils in semiarid environments through the recycling of organic matter. Appl. Environ. Soil Sci. 1–10.

Nancharaiah, Y. V, Lens, P.N.L., 2015b. Selenium biomineralization for biotechnological applications. Trends Biotechnol. 33, 323–330.

Nancharaiah, Y. V, Lens, P.N.L., 2015a. The ecology and biotechnology of selenium respiring bacteria. Microbiol. Mol. Biol. Rev. 79, 61–80.

Nancharaiah, Y. V, Mohan, S.V., Lens, P.N.L., 2016. Biological and bioelectrochemical recovery of critical and scarce metals. Trends Biotechnol. 34, 137–155.

Parkman, H., Hultberg, H., 2002. Occurence and effects of selenium in the environment - a literature review. IVL-rapport B1486, Swedish Environmental Research Institute, Göteborg.

Perkins, W.T., 2011. Extreme selenium and tellurium contamination in soils - An eighty year-old industrial legacy surrounding a Ni refinery in the Swansea Valley. Sci. Total Environ. 412–413, 162–169.

Ramos-Ruiz, A., Field, J.A., Wilkening, J. V., Sierra-Alvarez, R., 2016. Recovery of elemental tellurium nanoparticles by the reduction of tellurium oxyanions in a methanogenic microbial consortium. Environ. Sci. Technol. 50, 1492–1500.

Sura-de Jong, M., Reynolds, R.J.B., Richterova, K., Musilova, L., Staicu, L.C., Chocholata, I., Cappa, J.J., Taghavi, S., van der Lelie, D., Frantik, T., Dolinova, I., Strejcek, M., Cochran, A.T., Lovecka, P., Pilon-Smits, E. A. H., 2015. Selenium hyperaccumulators harbor a diverse endophytic bacterial community characterized by high selenium resistance and plant growth promoting properties. Front. Plant Sci. 6, 1–17.

Tan, L.C., Nancharaiah, Y. V, van Hullebusch, E.D., Lens, P.N.L., 2016. Selenium: Environmental significance, pollution, and biological treatment technologies. Biotechnol. Adv. 34, 886–907.

Tinggi, U., 2003. Essentiality and toxicity of selenium and its status in Australia : a review.

Toxicol. Lett. 137, 103–110.

Yasin, M., El Mehdawi, A.F., Jahn, C.E., Anwar, A., Turner, M.F.S., Faisal, M., Pilon-Smits, E. A. H., 2014. Seleniferous soils as a source for production of selenium-enriched foods and potential of bacteria to enhance plant selenium uptake. Plant Soil 386, 385–394.

Biography

Shrutika Laxmikant Wadgaonkar was born on June 29, 1986 at Aurangabad, Maharashtra, India. Shrutika did her bachelor studies (BSc) in biotechnology at the University of Mumbai and master studies (MSc) in biotechnology at the Dr. Babasaheb Ambedkar Marathwada University. She was qualified for the National Eligibility Test conducted jointly by the University Grants Commission and the Council for Scientific and Industrial Research (Lectureship) in December 2008. Upon graduation, Shrutika joined as 'Research Assistant' at the Centre for DNA Fingerprinting and Diagnostics, Hyderabad (India),  where she worked on the project entitled 'Screening and isolation of rpoB mutants in *E. coli* defective in transcription termination', after which she joined as 'Research Fellow' at the Department of Environmental Science, University of Mumbai, Mumbai (India) where she worked on the project entitled 'Bioremediation of dye stuff effluent compounds in a sequence bioreactor and metagenomic study of rhizosphere'. In 2014, Shrutika started her PhD program at UNESCO IHE – Institute for Water Education, Delft (the Netherlands), as part of an Erasmus Mundus Joint Doctorate Program on Environmental Technologies for Contaminated Solids, Soils and Sediments (ETeCoS^3). During her PhD, she also carried out her research at the Helmholtz institute for Environmental Research-UFZ, Leipzig (Germany) and University of Federico II, Naples (Italy). Shrutika also worked at the University of Saarland, Saarbrucken (Germany) during a short term scientific mission (STSM, COST Action ES1302). Her research was mainly focused on the development of a technology for the remediation of seleniferous soils/sediments and to explore microbial reduction of selenium oxyanions under different respiration conditions and bioreactor configurations.

Publications

- Ohlbaum M, **Wadgaonkar SL***, van Bruggen H, Nancharaiah YV, Esposito G, Lens PNL. (2018) Phytoremediation of seleniferous soil leachate using the aquatic plants *Lemna minor* and *Egeria densa*. Ecol. Eng. (*Accepted*). DOI: 10.1016/j.ecoleng.2018.06.013 (***corresponding author**)

- **Wadgaonkar SL**, Ferraro A, Nancharaiah YV, Dhillon KS, Fabbricino M, Esposito G, Lens PNL. (2018) *In situ* and *ex situ* bioremediation of seleniferous soils from Northwestern India. J. Soils Sediments (*Accepted*). DOI: 10.1007/s11368-018-2055-7

- **Wadgaonkar SL**, Mal J, Nancharaiah YV, Maheshwari N, Esposito G, Lens PNL. (2018) Formation of Se(0), Te(0) and Se(0)-Te(0) nanostructures during simultaneous bioreduction of selenite and tellurite in a UASB reactor. Appl. Microbiol. Biotechnol. 102(6): 2899-2911. DOI: 10.1007/s00253-018-8781-3

- **Wadgaonkar SL**, Nancharaiah YV, Esposito G, Lens PNL (2018) Environmental impact and bioremediation of seleniferous soils and sediments. Crit. Rev. Biotechnol. 38(6): 941-956. DOI: 10.1080/07388551.2017.1420623

- **Wadgaonkar SL**, Nancharaiah YV, Jacob C, Esposito G, Lens PNL. Selenate reduction by *Delftia lacustris* under aerobic conditions. Microbial Biotechnol. (*under review*)

- **Wadgaonkar SL**, Ferraro A, Race M, Nancharaiah YV, Dhillon KS, Fabbricino M, Esposito G, Lens PNL. Optimisation of soil washing to reduce selenium level of seleniferous soil from Punjab, Northwestern India. J. Environ. Qual. (*under review*).

- Fulekar MH, **Wadgaonkar SL**, Singh A (2013) Decolourization of dye compounds by selected bacterial strains isolated from dyestuff industrial area. Int. J. Adv. Res. Technol. 2(7):182-192.

- Fulekar MH, **Wadgaonkar SL**, Singh A (2013) Decolourization of dye compounds by selected bacterial strains isolated from dyestuff industrial area. Int. J. Adv. Res. Technol. 2(7):182-192.

Conferences

- **Wadgaonkar SL**, Nancharaiah YV, Esposito G, Lens PNL (2015) Selenate reduction by bacterial strains isolated from anaerobic granular sludge. 4th International Conference on Research Frontiers in Chalcogen Cycle Science & Technology (G16), UNESCO IHE, Delft (the Netherlands).

- Pathak P, **Wadgaonkar SL**, Fulekar MH (2012) Ecological Remediation of Persistent Dye Compounds in Soil Water Environment. Proceedings - Strategies for Mitigation of Environmental Degradation and Climate Change, Hisar, Haryana, India, pp 94-100.

D I P L O M A

For specialised PhD training

The Netherlands Research School for the
Socio-Economic and Natural Sciences of the Environment
(SENSE) declares that

Shrutika Laxmikant Wadgaonkar

born on 29 June 1986 in Aurangabad, India

has successfully fulfilled all requirements of the
Educational Programme of SENSE.

Delft, 18 December 2017

the Chairman of the SENSE board

Prof. dr. Huub Rijnaarts

the SENSE Director of Education

Dr. Ad van Dommelen

KONINKLIJKE NEDERLANDSE
AKADEMIE VAN WETENSCHAPPEN

The SENSE Research School declares that Ms Shrutika Laxmikant Wadgaonkar has successfully fulfilled all requirements of the Educational PhD Programme of SENSE with a work load of 51 EC, including the following activities:

SENSE PhD Courses

- o Anaerobic wastewater treatment (2015)
- o Environmental research in context (2015)
- o Research in context activity: 'Co-organizing programme and abstract book of 4th International Conference on Research Frontiers in Chalcogen Cycle Science and Technology' (2015)

Other PhD and Advanced MSc Courses

- o Contaminated soil and remediation, Paris-Est University (2015)
- o Contaminated soil, Paris-Est University (2015)
- o Advanced Biological Waste-to-Energy Technologies , UNESCO-IHE (2016)
- o Contaminated sediments – characterization and remediation, UNESCO-IHE (2016)
- o Mathematical models in environmental technologies, Entrepreneurship and innovation, and How to write successful proposal in Horizon 2020, University of Cassino (2017)
- o Biological treatment of solid waste, University of Cassino (2017)

External training at a foreign research institute

- o Molecular analysis: Automated Ribosomal Intergenic Spacer Analysis (ARISA) and cloning technique, University of Suor Orsola Benincasa of Naples (2017)

Management and Didactic Skills Training

- o Assisting microbiology laboratory session of MSc programme 'Environmental science and urban water supply' (2014-2016)
- o Supervising MSc student with thesis entitled 'Phytoremediation of seleniferous soil washing effluents by free floating aquatic plants, *Lemna minor* and *Egeria densa*' (2017)

Oral Presentations

- o *Removal and Bioreduction of Selenium Oxyanions from Seleniferous soils.* PhD Symposium: Integrating Research Water Sector, 28-29 September 2015, Delft, The Netherlands
- o *Optimisation of soil washing for Seleniferous soil from North-west India and biological treatment of soil leachate.* PhD Symposium: Climate Extremes and Water Management Challenge, 2-3 October 2017, Delft, The Netherlands

SENSE Coordinator PhD Education

Dr. Monique Gulickx